Peter F. Stiller

Automorphic Forms and the Picard Number of an Elliptic Surface

Aspects of Mathematics

Aspekte der Mathematik

Editor: Klas Diederich

The texts published in this series are intended for graduate students and all mathematicians who wish to broaden their research horizons or who simply want to get a better idea of what is going on in a given field. They are introductions to areas close to modern research at a high level and prepare the reader for a better understanding of research papers. Many of the books can also be used to supplement graduate course programs.

The series comprises two sub-series, one with English texts only and the other in German.

Peter F. Stiller

Automorphic Forms and the Picard Number of an Elliptic Surface

Friedr. Vieweg & Sohn   Braunschweig/Wiesbaden

Dr. *Peter F. Stiller* is Associate Professor of Mathematics at Texas A & M University, College Station, Texas 77843, USA.

1984

Procuded by IVD, Walluf b. Wiesbaden

ISBN 978-3-322-90710-3   ISBN 978-3-322-90708-0 (eBook)
DOI 10.1007/978-3-322-90708-0

INDEX

The author wishes to express his thanks to the Sonderforschungsbereich of the Universität Bonn, West Germany for its invitation and financial support during all of 1981, and for providing a stimulating and rewarding atmosphere in which to work. Parts of this manuscript were also written during the 1982-1983 academic year while the author was a guest of the Institut des Hautes Études Scientifiques. Lastly, thanks must go to the referee for his many thoughtful comments which have improved the presentation of these results. Any remaining errors are the sole responsibility of the author.

INTRODUCTION

In studying an algebraic surface E, which we assume is non-singular and projective over the field of complex numbers $\mathbb{C}$, it is natural to study the curves on this surface. In order to do this one introduces various equivalence relations on the group of divisors (cycles of codimension one). One such relation is algebraic equivalence and we denote by NS(E) the group of divisors modulo algebraic equivalence which is called the Néron-Severi group of the surface E. This is known to be a finitely generated abelian group which can be regarded naturally as a subgroup of $H^2(E,\mathbb{Z})$. The rank of NS(E) will be denoted ρ and is known as the Picard number of E.

Every divisor determines a cohomology class in $H^2(E,\mathbb{C})$ which is of type (1,1), that is to say a class in $H^1(E,\Omega_E^1)$ which can be viewed as a subspace of $H^2(E,\mathbb{C})$ via the Hodge decomposition. The Hodge Conjecture asserts in general that every rational cohomology class of type (p,p) is algebraic. In our case this is the Lefschetz Theorem on (1,1)-classes: Every cohomology class

$$\gamma \in H^1(E,\Omega_E^1) \cap H^2(E,\mathbb{Z})$$

is the class associated to some divisor. Here we are writing $H^2(E,\mathbb{Z})$ for
its image under the natural mapping into $H^2(E,\mathbb{C})$. Thus NS(E) modulo
torsion is $H^1(E,\Omega_E^1) \cap H^2(E,\mathbb{Z})$ and ρ measures the algebraic part of the
cohomology.

In this paper we shall be interested in a certain class of elliptic
surfaces and in the problem of extracting information about ρ directly
from the Gauss–Manin connection for such families of elliptic curves. We
shall develop a method to determine the Picard number (or more precisely,
to establish the existence of sections of the elliptic surface E over its
base curve X) by reducing the problem to the computation of certian mixed
hypergeometric and abelian integrals on a punctured sphere and we shall
interpret these integrals in numerous interesting ways — namely, as the
periods of an inhomogeneous differential equation; as the periods of a
generalized automorphic form; in terms of the monodromy of the differential
equation satisfied by a "normal function"; in terms of the special values
of certain Dirichlet series; as extension class data for a locally split
extension of flat vector bundles (see Stiller [36]); and lastly from the
standpoint of the representation theory of a certain parabolic subgroup of
SL_3 (see Vilenkin [35]). Our primary tools will be the Gauss–Manin
connection associated to the surface expressed as a homogeneous second
order linear differential equation Λ with regular singular points on the
base curve X and an inhomogeneous de Rham cohomology created from certain
inhomogeneous differential equations formed with this operator Λ (along
the lines of Atiyah and Hodge [41]: locally exact modulo exact).

Essential to understanding our point of view is the Eichler–Hoyt
correspondence between normal functions and generalized automorphic forms,
and the techniques inaugurated by Hoyt in [12]. In addition it is useful

to have some knowledge of hypergeometric integrals. The most complete up-to-date source is Exton [40]. Our integrals are the analogues of those that arise in the theory of classical automorphic forms (see Stiller [36] and Chowla and Gross [44]).

In all that follows E will be an elliptic surface having a global section over its base curve X. It is assumed throughout that the functional invariant J is non-constant and that E has no exceptional curves of the first kind in its fibers. We shall utilize the general theory of elliptic surfaces due to Kodaira [16] and [17], and properties of the Gauss-Manin connection associated with an elliptic surface (Stiller [28] and [29]).

Letting $K(X)$ be the function field of the base curve X, we denote by $E^{gen}(K(X))$ the group of $K(X)$-rational points on the generic fiber E^{gen} of E over X which is an elliptic curve over $K(X)$. By the Mordell-Weil Theorem, $E^{gen}(K(X))$ is a finitely generated abelian group whose rank we denote by r. We refer the reader to Shioda [27] for a description of $NS(E)$ and its relation to the group $E^{gen}(K(X))$. In particular recall

$$\rho = r + 2 + \sum_{\nu} (m_{\nu} - 1)$$

where ν is an index running over the singular fibers of the elliptic surface and m_{ν} is the number of irreducible components of the fiber. We shall focus on the rank r which is also the rank of the group of sections of E over X, but as the above formula shows, any method to determine r allows us to determine ρ.

In the first section we shall discuss differential equations in general and certain automorphic forms that can be attached to them. We then focus attention on a special class of differential equations known as K-equations which are the relevant ones to consider when dealing with elliptic surfaces. In this case the associated automorphic forms are especially interesting and we can interpret the periods of these forms in a number of useful ways as mentioned above. In the third part we compare some bounds obtained on ρ by Stiller [29] and Hoyt [12], and give proofs of some of Hoyt's results by direct computation of the automorphic forms involved; including a generalization of a result in Shioda [27] identifying $H^0(E,\Omega_E^2)$ with a certain space of automorphic forms. Part four relates these results, the differential equations, and the automorphic forms constructed from them to the Hodge theory of E over X. As a sidelight, we give some applications to determining the torsion in $E^{gen}(K(X))$. Finally, in part five we give some of the interpretations of the periods and a method for determining the Picard number — illustrating it with some examples. That these examples and computations involve a great deal of the theory of special functions is easily understandable in light of the analogy with classical automorphic forms. For what we are really doing (in one interpretation) is finding the special values of certain Dirichlet series, and it thereby comes as no surprise when the values that we compute are ratios of products of values of the gamma function. (For some useful rationality properties of values of the gamma function see the appendix by Koblitz and Ogus in Deligne [42].) In fact our methods can be adapted to the classical case of the special values in the critical strip of the Dirichlet series associated to a cusp form of weight greater than two (Stiller [36]). In this case one can sometimes make use of the formulas of Damerell and Chowla-Selberg — for example when the cusp form is attached to a grossen-character. In all the examples we do, the integrands used to compute the periods resemble those of the classical beta functions times a hypergeometric function and are those which Exton [40] refers to as integrals of Euler type. These represent parts of weight two (specifically the beta function appears in the periods of the Fermat curves, see Gross [37]) and weight one (carrying the representation) respectively in the generalized automorphic form of weight three.

The appearance of values of the gamma function etc. is also not surprising when one realizes that from another point of view we are computing the monodromy representations of a class of third order linear homogeneous differential equations (the differential equations satisfied by the "normal functions") -- which in itself is an interesting problem. We obain a number of examples of third order equations with $SL_3(\mathbb{Z})$ monodromy in this way, and when we consider families of surfaces we can see how the monodromy of certain equations (with $SL_3(\mathbb{C})$ monodromy) depends on the parameters (see Bateman [1] for the case of the classical hypergeometric equation). We are also able in some cases to identify the third order equation satisfied by some normal function as a composite involving a generalized hypergeometric $_3F_2$.

In short, our various interpretations lead to a relationship between the existence of algebraic cycles, rationality properties of certain monodromy representations, and rationality properties of special values (along the lines of the rationality results of Manin [45] and Razar [46] for the special values of the Dirichlet series attached to classical automorphic forms which are common eigenfunctions of the Hecke algebra). It is the link betwen these classical problems that has inspired our interest in the problems considered here.

PART I. DIFFERENTIAL EQUATIONS

§1. <u>Generalities.</u>

Let X be a projective non-singular curve over the field of complex numbers $\mathbb{C}$. We consider a homogeneous second order linear differential equation

$$1) \qquad \Lambda f = \frac{d^2 f}{dx^2} + P \frac{df}{dx} + Q f = 0$$

where f is an unknown function, $P, Q \in K(X)$ the function field of X, and $x \in K(X)$ is non-constant and serves as a local parameter almost everywhere on X. We shall always assume Λ has regular singular points.

We can find a Zariski open set X_0 on which P, Q are regular and x is a local parameter. Pick a base point $x_0 \in X_0$. In a neighborhood of x_0 we can find two holomorphic solutions ω_1, ω_2 of Λ which form a basis for the space of solutions near x_0. That is, every solution is of the form $c_1 \omega_1 + c_2 \omega_2$ in a neighborhood of x_0 with $c_1, c_2 \in \mathbb{C}$.

The pair ω_1, ω_2 can be analytically continued around any closed path $\gamma \in \pi_1(X_0, x_0)$ and the result will be a certain linear combination of the original pair. We denote this by

$$\begin{pmatrix} \omega_1 \\ \omega_2 \end{pmatrix} \rightarrow M_\gamma \begin{pmatrix} \omega_1 \\ \omega_2 \end{pmatrix} \qquad M_\gamma \in GL_2(\mathbb{C})$$

that is,

$$\begin{aligned} \omega_1 &\longrightarrow a_\gamma \omega_1 + b_\gamma \omega_2 \\ \omega_2 &\longrightarrow c_\gamma \omega_1 + d_\gamma \omega_2 \end{aligned} \qquad \begin{pmatrix} a_\gamma & b_\gamma \\ c_\gamma & d_\gamma \end{pmatrix} = M_\gamma .$$

This is known as the <u>monodromy representation.</u>

We shall assume X_0 is uniformized by the complex upper-half-plane $\hbar = \{z \in \mathbb{C} \text{ s.t. } \operatorname{Im} z > 0\}$. Let $\pi: \hbar, z_0 \to X_0, x_0$ be the universal covering map with $\pi(z_0) = x_0$ a selected base point. We can lift ω_1, ω_2 to $\hbar$ with the chosen branches of ω_1, ω_2 at x_0 being based at $z_0 \in \hbar$ so as to give a well-defined map $\omega = \omega_1/\omega_2$ on $\hbar$:

$$
\begin{array}{ccc}
z_0, \hbar & \xrightarrow{\ \omega\ } & \mathbb{P}^1_{\mathbb{C}} \\
{\scriptstyle \pi}\big\downarrow & & \\
x_0, X_0 & &
\end{array}
\qquad \cdot
$$

We call $\omega = \omega_1/\omega_2$ the <u>period map</u>. If $\overline{\Gamma} \subset \mathrm{PSL}_2(\mathbb{R})$ represents the group of covering translations (which is isomorphic to $\pi_1(X_0, x_0)$) we have for $\gamma \in \overline{\Gamma}$

$$\omega(\gamma z) = M_\gamma \omega(z)$$

where M_γ acts via linear fractional transformation.

We shall need some terminology to describe the behaviour of M_γ, ω_1, ω_2 and ω at or around the missing points in $S = X - X_0$. For $x \in S$, if $M_\gamma = \begin{pmatrix} 1 & 0 \\ 0 & 1 \end{pmatrix}$ for a small loop around x, that is if there is no local monodromy around x, we call x a <u>cosingular point.</u> At such a point ω_1, ω_2, and ω are single-valued but perhaps meromorphic. (By assuming that Λ always has regular singular points we have ruled out an essential singularity in this case.)

We note that it is often desirable to include these cosingular points among the points of X_0 and later on, depending on the specific problem under consideration, we may actually do so. Of course one will need to check that the constructions make sense on this larger open set. For example, the monodromy representation and the period map are both defined in this extended setting.

Points where the local monodromy is non-trivial will be called <u>true</u>
<u>singular points</u> and said to be parabolic, elliptic, hyperbolic or
loxodromic when the local monodromy is of that type respectively (see
Shimura [25]).

For any non-zero rational function $g \in K(X)$ we denote by Λ_g the
"twisted" equation whose solutions are g times those for Λ. Thus $g\omega_1$,
$g\omega_2$ gives a basis as before and we see that both the monodromy representation and the period map are unchanged. Although one changes the set of
cosingular points by this process, it is generally quite harmless.

For further details the reader can consult Ince [14], Poincaré [23],
Picard [22], Deligne [5] or Griffiths [10].

§2. <u>Inhomogeneous equations.</u>

In what follows, we shall be interested in the inhomogeneous equations
formed from the operator Λ given in 1) above. Thus for $Z \in K(X)$
consider

2) $\Lambda f = Z$ f   unknown

Choose $X_0 \subset X$ as above but make it small enough so that Z is also
regular on X_0 and select a base point $x_0 \in X_0$ together with branches of
ω_1 and ω_2 at x_0 as before.

<u>Theorem I.2.1</u>: $f = \left(\int_{x_0}^{x} \dfrac{-\omega_2 Z}{W} dx + c_1 \right)\omega_1 + \left(\int_{x_0}^{x} \dfrac{\omega_1 Z}{W} dx + c_2 \right)\omega_2$

as a multivalued holomorphic function on X_0. Here c_1, $c_2 \in \mathbb{C}$ and
$W = \omega_1 \dfrac{d\omega_2}{dx} - \omega_2 \dfrac{d\omega_1}{dx}$ is the Wronskian of Λ. Note that $W = e^{-\int Pdx}$ up to
a non-zero constant multiple so that on X_0 it is invertible. We have

9

also somewhat abused notation in using x to denote both the parameter

$x \in K(X)$ and a variable point $x \in X_0$. □

Proof: Apply variation of parameters to 2).

Note $\Lambda_g(gf) = gZ$ for $g \in K(X)$ non-zero. This shows that the
expressions $\left(\int_{x_0}^{x} \frac{-\omega_2 Z}{W} dx + c_1\right)$ and $\left(\int_{x_0}^{x} \frac{\omega_1 Z}{W} dx + c_2\right)$ are "invariant" under
twist if we define the twist of an inhomogeneous equation by replacing Z
with gZ and Λ with Λ_g. The invariance follows as the solutions of Λ_g
are $g\omega_1$, $g\omega_2$ and the new Wronskian is g^2W where W is the Wronskian for
Λ. This remark will be useful later when we look at elliptic surfaces.

We also wish to observe the effect of a parameter change. If we
replace the parameter $x \in K(X)$ by $t \in K(X)$ the new inhomogeneous
equation is

$$\Lambda^{new}f = \frac{d^2f}{dt^2} + \left(P \frac{dx}{dt} - \frac{d}{dt} \log \frac{dx}{dt}\frac{df}{dt}\right) + Q\left(\frac{dx}{dt}\right)^2 f = 0$$

where P, Q are the original coefficients; see 1) above. Moreover we have
for the inhomogeneous case:

$$\Lambda^{new}f = Z\left(\frac{dx}{dt}\right)^2$$

where f is the original inhomogeneous solution; see 2) above and Theorem
I.2.1. Note that the Wronskian W, which clearly depends on a choice of
derivation $\frac{d}{dx}$, becomes $W \frac{dx}{dt}$ when we change parameter. We can thus see
directly from the formula in Theorem I.2.1 that the expressions

$$\frac{-\omega_2 Z}{W} dx \quad \text{and} \quad \frac{\omega_1 Z}{W} dx$$

are naturally independent of parameter since

$$\frac{-\omega_2 z \left(\frac{dx}{dt}\right)^2}{W\left(\frac{dx}{dt}\right)} \, dt \;=\; \frac{-\omega_2 z}{W} \, dx$$

and

$$\frac{\omega_1 z \left(\frac{dx}{dt}\right)^2}{W\left(\frac{dx}{dt}\right)} \, dt \;=\; \frac{\omega_1 z}{W} \, dx.$$

We now return to consideration of the inhomogeneous equation.

Clearly we can continue the solution f of the inhomogeneous equation in an unrestricted way throughout X_0. For $\gamma \in \pi_1(X_0, x_0)$ we have

$$f \longrightarrow f + m_\gamma \omega_1 + n_\gamma \omega_2 \qquad m_\gamma, \; n_\gamma \in \mathbb{C}$$

because any two solutions of the inhomogeneous equation differ by a solution of the homogeneous equation.

Lifting f to the upper-half-plane viewed as the universal cover of X_0 as in §1 above, we have a holomorphic single-valued function which we also denote by f and which transforms as

$$f(\gamma z) = f(z) + m_\gamma \omega_1(z) + n_\gamma \omega_2(z), \qquad m_\gamma, \; n_\gamma \in \mathbb{C}$$

for $z \in \mathcal{h}$ and $\gamma \in \overline{\Gamma} \subset PSL_2(\mathbb{R})$ corresponding to $\gamma \in \pi_1(X_0, x_0)$. Note that near the missing points f satisfies growth conditions similar to those for ω_1, ω_2 which are solutions of Λ an equation with regular singular points.

§3. Automorphic forms.

In order to simplify the discussion in this section we shall make a number of assumptions which in many cases can be relaxed or omitted entirely. Later, when dealing with elliptic surfaces, all the assumptions will hold automatically.

First we shall assume that the global monodromy of our differential equation Λ is in $SL_2(\mathbb{C})$. This is equivalent to the Wronskian W being in $K(X)$. We will fix a finite set of points $S \subset X$ so that on the Zariski open subset $X_0 = X - S$ the equation Λ will be holomorphic, and will possess a basis for its space of solutions ω_1, ω_2 which are holomorphic <u>non-vanishing</u> multivalued functions on X_0. We shall assume $x \in K(X)$ is a good lcoal parameter at all points of X_0 (although this is not particularly essential). Finally we shall only consider $Z \in L(*S) = \{h \in K(X)$ such that the poles of h are contained in $S\}$.

As before we consider the inhomogeneous equation

$$\Lambda f = Z$$

and the solution as in Theorem I.2.1,

$$f = \left(\int_{x_0}^x \frac{-\omega_2 Z}{W}\, dx + c_1 \right)\omega_1 + \left(\int_{x_0}^x \frac{\omega_1 Z}{W}\, dx + c_2 \right)\omega_2$$

where $x_0 \in X_0$ is a base point and ω_1, ω_2 are fixed branches at x_0. Because of our assumptions f is holomorphic multivalued on X_0 as is the function

$$F = \frac{f}{\omega_2} = \left(\int_{x_0}^x \frac{-\omega_2 Z}{W}\, dx + c_1 \right)\omega + \left(\int_{x_0}^x \frac{\omega_1 Z}{W}\, dx + c_2 \right).$$

We define $F(z)$, a holomorphic single-valued function on the upper-half-plane $\mathcal{h}$, to be the lift of F via the universal cover $\pi: \mathcal{h}$, $z_0 \to X_0$, x_0 as in section §1 above.

Note that for $\gamma \in \overline{\Gamma} \subset PSL_2(\mathbb{R})$ corresponding to $\gamma \in \pi_1(X_0, x_0)$ we have:

3)
$$F(\gamma z) = (c_\gamma \omega(z) + d_\gamma)^{-1}(F(z) + m_\gamma \omega(z) + n_\gamma)$$

where m_γ , $n_\gamma \in \mathbb{C}$ are called the **periods** of F around γ , and $\begin{pmatrix} a_\gamma & b_\gamma \\ c_\gamma & d_\gamma \end{pmatrix} \in$
$SL_2(\mathbb{C})$ is the monodromy matrix for the continuation of ω_1 , ω_2 around the corresponding path. As always $\omega = \omega_1/\omega_2$ which in this case is the period map viewed as a function on $\mathcal{h}$. Again we remark that regularity imposes growth conditions on F near the points of S.

By analogy with Eichler [9] and Hoyt[12] for each holomorphic F on the upper-half-plane $\mathcal{h}$ satisfying 3) there is a holomorphic function g on $\mathcal{h}$ satisfying

$$g(\gamma z) = (c_\gamma \omega(z) + d_\gamma)^3 g(z)$$

$z \in \mathcal{h}$, $\gamma \in \overline{\Gamma} \subset PSL_2(\mathbb{R})$ and $\begin{pmatrix} a_\gamma & b_\gamma \\ c_\gamma & d_\gamma \end{pmatrix}$ the monodromy as before. These functions are related by the formulas:

4)
$$F(z) = \int_{z_0}^{z} g(t)(\omega(z) - \omega(t))d\omega(t) + c_1 \omega(z) + c_2$$

5)
$$g(z) = \frac{1}{\pi i} \oint F(t)(\omega(t) - \omega(z))^{-3} d\omega(t)$$

where the path is taken around the boundary of a small disc about $z \in \mathcal{h}$.

__Theorem I.3.1.__ $g(z) = \dfrac{\omega_2(z)^3 Z(z)}{W(z)^2}.$

__Proof.__ We analyze the situation locally near x_0. Here

$$F(x) = f(x)/\omega_2(x) = \left(\int_{x_0}^{x} \frac{-\omega_2 Z}{W} dt + c_1\right)\omega(x) + \left(\int_{x_0}^{x} \frac{\omega_1 Z}{W} dt + c_2\right).$$

Note that we have replaced the parameter x by t to avoid confusion with the point $x \in X_0$. Think of t as $x - x(x_0)$ being a local coordinate at x_0. Let U be a small neighborhood of x_0 and denote $\omega(U) \subset \mathbb{C}$ by V so that we have:

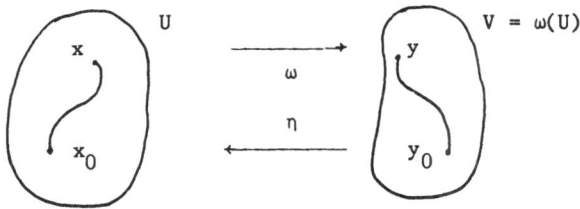

where η is a local inverse to ω. The formulas 4) and 5) are equivalent to

$$(g \circ \eta)(y) = \frac{d^2}{dy^2} (F \circ \eta)(y).$$

We compute

$$(F \circ \eta)(y) = F(\eta(y)) = \left(\int_{\eta(y_0)}^{\eta(y)} \frac{-\omega_2(t)Z(t)}{W(t)} dt + c_1\right)\omega(\eta(y))$$

$$+ \left(\int_{\eta(y_0)}^{\eta(y)} \frac{\omega_1(t)Z(t)}{W(t)} dt + c_2\right).$$

Note $\omega(\eta(y)) = y$. We now change variables in the integral setting $t = \eta(s)$ or $\omega(t) = s$. Thus

$$F(\eta(y)) = \left(\int_{y_0}^{y} \frac{-\omega_2(\eta(s))Z(\eta(s))}{W(\eta(s))} \eta'(s)ds + c_1\right)y$$

$$+ \left(\int_{y_0}^{y} \frac{\omega_1(\eta(s))Z(\eta(s))}{W(\eta(s))} \eta'(s)ds + c_2\right).$$

We now differentiate with respect to y. First

$$\frac{d}{dy} (F \circ \eta)(y) = \frac{-\omega_2(\eta(y))Z(\eta(y))}{W(\eta(y))} \eta'(y)y$$

$$+ \left(\int_{y_0}^{y} \frac{-\omega_2(\eta(s))Z(\eta(s))}{W(\eta(s))} \eta'(s)ds + c_1\right)$$

$$+ \frac{\omega_1(\eta(y))Z(\eta(y))}{W(\eta(y))} \eta'(y)$$

But the first and third term add to zero as

$$\frac{d\eta}{dy} = \frac{1}{\frac{d\omega}{dx}(\eta(y))} = \frac{\omega_2(\eta(y))^2}{-W(\eta(y))} .$$

because

$$\frac{d\omega}{dx} = \frac{\omega_2\omega_1' - \omega_1\omega_2'}{\omega_2^2} = \frac{-W}{\omega_2^2} ,$$

and

$$\frac{d\eta}{dy} \cdot y = \frac{\omega_2(\eta(y))\omega_1(\eta(y))}{-W(\eta(y))}$$

because

$$y = \omega(\eta(y)) = \omega_1(\eta(y))/\omega_2(\eta(y)).$$

Thus

$$(g \circ \eta)(y) = \frac{d^2}{dy^2} (F \circ \eta)(y) = \frac{-\omega_2(\eta(y))Z(\eta(y))}{W(\eta(y))} \eta'(y)$$

$$= \frac{\omega_2(\eta(y))^3 Z(\eta(y))}{W^2(\eta(y))}$$

where $\eta'(y)$ is as determined above. The result follows by setting
$y = \omega(x)$, etc.   □

We wish to make two remarks. The first is in regard to the invariance
of $g = \dfrac{\omega_2{}^3 Z}{W^2}$. This "automorphic form" is completely independent of the
choice of a particular solution to $\Lambda f = Z$, i.e. it is independent of c_1
and c_2, as well as the choice of base point $x_0 \in X_0$ (provded we choose
the branches of ω_1, ω_2 appropriately at the new base point). Moreover it
is independent of the parameter chosen since as remarked above Z becomes
$Z\left(\dfrac{dx}{dt}\right)^2$ and W^2 becomes $W^2\left(\dfrac{dx}{dt}\right)^2$. The expression for g is also
invariant of "twist" of Λ by $h \in K(X)$ to Λ_h provided we replace Z
with hZ. We will see that this replacement has a natural description in
the elliptic surface case. Thus g is dependent only on the choice of
ω_1, ω_2. (Note that if we fix ω_1, ω_2 but choose other initial branches
then g changes by the natural action of $\overline{\Gamma}$ via the monodromy
representation.)

Next we remark that the above construction can be carried through on a
fixed X_0 where Λ has no true singular points but possibly cosingular
points and where Z has possible poles. F will then be holomorphic
multivalued on $\hbar$ minus the orbits of a finite set of points. We can
still define g and the above theorem holds, however g may well have
poles on $\hbar$ although it will be single-valued.

Thus to each rational function $Z \in L(*S)$ (or $\in K(X)$) we can
associate a holomorphic (or meromorphic) function g on $\hbar$ viewed as the
universal cover of X_0 which satisfies the transformation rule:

6) $\qquad g(\gamma z) = (c_\gamma \omega(z) + d_\gamma)^3 g(z) \qquad\qquad \gamma \in \overline{\Gamma} \subset PSL_2(\mathbb{R})$

$$\begin{pmatrix} a_\gamma & b_\gamma \\ c_\gamma & d_\gamma \end{pmatrix} \quad \text{the monodromy.}$$

We remark again that because of regular singularities g has reasonable growth at the cusps of $\overline{\Gamma}$. Thus we obtain an "automorphic form" of weight 3.

In later parts of this work we will discuss a number of specific examples, including some cases where we in fact get classical automorphic forms and we will examine under what conditions they give holomorphic automorphic forms and/or cusp forms in the usual sense. Note that ω_2 itself transforms as

$$\omega_2(\gamma z) = (c_\gamma \omega(z) + d_\gamma)\omega_2(z)$$

so it behaves like a form of weight 1. In all cases of interest Λ will have $SL_2(\mathbb{R})$ monodromy with the projective monodromy group a Fuchsian group of the first kind in $PSL_2(\mathbb{R})$, and the equation will have positivity, that is it will have a basis of solutions ω_1, ω_2 with $Im(\omega_1/\omega_2) > 0$.

We shall mention one example before going on. Let Λ be an equation which uniformizes a curve X (possibly not complete), with universal cover $\mathcal{h}$. Then the period map ω is the identity (up to the obvious action of $PSL_2(\mathbb{R})$) for an appropriate choice of ω_1, ω_2. In this case we get the usual automorphic forms of weight 3 (allowing poles), and ω_2 will be a form of weight one (see also Stiller [31]).

§4. Periods.

In this section we want to examine in greater detail the periods of the automorphic forms we have produced. Recall

$$f = \left(\int_{x_0}^x \frac{-\omega_2 Z}{W} dx + c_1\right)\omega_1 + \left(\int_{x_0}^x \frac{\omega_1 Z}{W} dx + c_2\right)\omega_2.$$

The periods then come from continuing f around a path $\gamma \in \pi_1(X_0, x_0)$:

$$f \;\to\; f + m_\gamma \omega_1 + n_\gamma \omega_2 \qquad m_\gamma,\; n_\gamma \in \mathbb{C}.$$

We will assume Z is regular on X_0 to avoid some complications but much of what we do here can be modified to account for poles of Z in X_0 (see Part II, §3). We set

$$\chi_1(x) = \int_{x_0}^{x} \frac{-\omega_2 Z}{W}\, dx \quad \text{and} \quad \chi_2(x) = \int_{x_0}^{x} \frac{\omega_1 Z}{W}\, dx.$$

We wish to examine how χ_1 continues around $\gamma \in \pi_1(X_0, x_0)$. The continuation of $\chi_1(x)$ is

$$(7) \quad \int_\gamma \frac{-\omega_2 Z}{W}\, dx + \int_{x_0}^{x} \frac{(-c_\gamma \omega_1 - d_\gamma \omega_2) Z}{W}\, dx = \int_\gamma \frac{-\omega_2 Z}{W}\, dx + -c_\gamma \chi_2(x) + d_\gamma \chi_1(x)$$

where the first integral is understood in the obvious fashion and is computed by continuation of our fixed branches of ω_1, ω_2 from x_0.

The second and third terms in (7) are the result of the multivalued nature of ω_2. Here $\begin{pmatrix} a_\gamma & b_\gamma \\ c_\gamma & d_\gamma \end{pmatrix}$ is the monodromy matrix. For $\chi_2(x)$ we get:

$$\int_\gamma \frac{\omega_1 Z}{W}\, dx + a_\gamma \chi_2(x) - b_\gamma \chi_1(x).$$

Thus:

<u>Proposition I.4.1</u>: $[\chi_1, \chi_2] \to [\chi_1, \chi_2] \left(M_\gamma^{-1}\right) + [\chi_1(\gamma), \chi_2(\gamma)]$

where $\chi_1(\gamma) = \int_\gamma \dfrac{-\omega_2 Z}{W}\, dx$, $\chi_2(\gamma) = \int_\gamma \dfrac{\omega_1 Z}{W}\, dx$. □

Now $f = ([\chi_1, \chi_2] + [c_1, c_2]) \begin{pmatrix} \omega_1 \\ \omega_2 \end{pmatrix}$ and under continuation around γ we get:

$$f + m_\gamma \omega_1 + n_\gamma \omega_2 = ([\chi_1, \chi_2] M_\gamma^{-1} + [\chi_1(\gamma), \chi_2(\gamma)] + [c_1, c_2]) M_\gamma \begin{pmatrix} \omega_1 \\ \omega_2 \end{pmatrix}$$

$$= f + ([\chi_1(\gamma), \chi_2(\gamma)] M_\gamma + [c_1, c_2](M_\gamma - I)) \begin{pmatrix} \omega_1 \\ \omega_2 \end{pmatrix}.$$

Thus:

<u>Proposition I.4.2</u>: The periods m_γ, n_γ of f (and so F and g) are:

$$[m_\gamma, n_\gamma] = [\chi_1(\gamma), \chi_2(\gamma)]M_\gamma + [c_1, c_2](M_\gamma - I).$$ □

These periods can be related to certain parabolic cohomology classes and we will discuss this more completely in the next part when the specific cases under consideration yield more particular results many of which hold in the more general setting of this section.

We give however one important result. We say f has <u>trivial periods</u> if there exists $[c_1', c_2'] \in \mathbb{C}^2$ such that the periods $[m_\gamma, n_\gamma]$ of f are $[c_1', c_2'](M_\gamma - I)$ for every path γ.

<u>Theorem I.4.3</u>: f has trivial periods iff the Z used in constructing f is $\Lambda Z'$ for some $Z' \in K(X)$. Thus the automorphic form $g = \dfrac{\omega_2^3 Z}{W}$ has trivial periods if and only if Z is as above.

<u>Proof</u>: f has trivial periods if and only if there exists $[c_1', c_2'] \in \mathbb{C}^2$ such that $\tilde{f} = f - c_1'\omega_1 - c_2'\omega_2$ has zero periods. This means $\tilde{f}$ is single-valued and because of growth conditions $\tilde{f} \in K(X)$. But $\Lambda\tilde{f} = \Lambda f = Z$ as Λ kills ω_1, ω_2, so Z is in the image of $\Lambda: K(X) \to K(X)$. □

Before we close this section, we also want to analyze the local behaviour of the periods $[m_\gamma, n_\gamma]$. By this we mean to consider a small loop γ around a missing point $s \in S = X - X_0$ as shown:

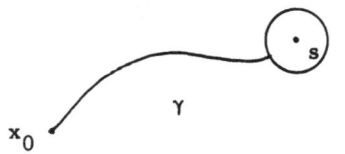

To do this we consider a local situation. Let h(z) be a multivalued holomorphic function in a neighborhood of the punctured unit disc, and consider for every $0 < \varepsilon \leq 1$ the path $\gamma_\varepsilon = \gamma_\varepsilon^{in} + C_\varepsilon + \gamma_\varepsilon^{out}$ where γ_ε^{in} is the path from -1 to $-\varepsilon$ along the negative real axis, C_ε is the circular path of radius ε centered at 0 traveled counterclockwise from $-\varepsilon$ to $-\varepsilon$, and γ_ε^{out} is $-\gamma_\varepsilon^{in}$ from $-\varepsilon$ to -1:

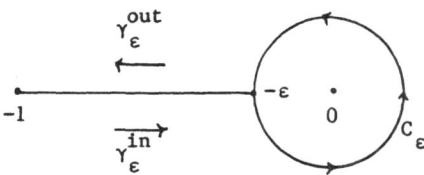

We fix a branch of h at -1 and try to analyze $\int_{\gamma_\varepsilon} h(z)dz$. We make one assumption -- "the regularity assumption" -- namely that there is an integer N such that in every sector about 0 and for every branch of h we have

$$\lim_{z \to 0} z^N h(z)$$

exists. We make a change of variables letting $z = e^w$, lifting $h(z)$ to a function of w analytic in a neighborhood of the left-half-plane $\text{Re } w \leq 0$. We base the branch of h chosen at -1 at the point $-i\pi$ in the w-plane. Letting $\varepsilon = e^r$ we have γ_ε corresponding to a path τ_r as shown:

and $\int_{\gamma_\varepsilon} h(z)dz$ becomes $\int_{\tau_r} h(e^w)e^w dw$. Using this one can show:

Proposition I.4.4: If N can be taken ≤ 1 above then $\lim\limits_{\varepsilon \to 0} \int_{C_\varepsilon} h(z)dz$ exists and is zero if N can be taken < 1. Moreover when N can be taken < 1 both $\lim\limits_{\varepsilon \to 0} \int_{\gamma_\varepsilon^{in}} h(z)dz$ and $\lim\limits_{\varepsilon \to 0} \int_{\gamma_\varepsilon^{out}} h(z)dz$ exist. □

We return now to the original question of the local behaviour of the periods $[m_\gamma, n_\gamma]$ for f. Recall

$$f = [(x_1, x_2) + (c_1, c_2)]\binom{\omega_1}{\omega_2}$$

and

$$[m_\gamma, n_\gamma] = [x_1(\gamma), x_2(\gamma)]M_\gamma + [c_1, c_2](M_\gamma - I).$$

Proposition I.4.5: $[m_{\gamma\tau}, n_{\gamma\tau}] = [m_\gamma, n_\gamma]M_\gamma + [m_\tau, n_\tau]$ where $\gamma\tau$ indicates first follow the path γ and then τ. □

An easy computation gives

Proposition I.4.6: $[x_1(\gamma\tau), x_2(\gamma\tau)] = [x_1(\gamma), x_2(\gamma)] + [x_1(\tau), x_2(\tau)]M_\gamma^{-1}$. □

Now we examine the continuation of x_1 around a path γ which is a simple loop around a missing point $s \in S = X - X_0$:

We write $\gamma = \gamma_1 + C + \gamma_1^{-1}$ where γ_1 traces the path up to the circular loop and C is the circular loop. We compute:

$$x_1(\gamma) = \int_{\gamma_1} \frac{-\omega_2 Z}{W} dx + \int_C \frac{-\omega_2 Z}{W} dx + \int_{\gamma_1^{-1}} \frac{(-c_{\gamma_1}\omega_1 - d_{\gamma_2}\omega_2)Z}{W} dx$$

$$= x_1(\gamma_1) + c_\gamma x_2(\gamma_1) + b_\gamma x_1(\gamma_1) + \int_C \frac{-\omega_2 Z}{W} dx$$

and likewise

$$\chi_2(\gamma) = \chi_2(\gamma_1) - a_\gamma \chi_2(\gamma_1) + b_\gamma \chi_1(\gamma_1) + \int_C \frac{\omega_1 Z}{W} \, dx.$$

Thus

<u>Proposition I.4.7</u>: $[\chi_1(\gamma),\ \chi_2(\gamma)] = [\chi_1(\gamma_1),\ \chi_2(\gamma_1)](I - M_\gamma^{-1})$

$$+ \left[\int_C \frac{-\omega_2 Z}{W} \quad \int_C \frac{\omega_1 Z}{W} \, dx \right]$$

and

$$[m_\gamma, n_\gamma] = \left[\int_C \frac{-\omega_2 Z}{W} \,,\ \int_C \frac{\omega_1 Z}{W} \right] M_\gamma + [(\chi_1(\gamma_1),\ \chi_2(\gamma_1)) + (c_1, c_2)](M_\gamma - I). \quad \square$$

We refer to

$$\int_C \frac{-\omega_2 Z}{W} \, dx \quad \text{and} \quad \int_C \frac{\omega_1 Z}{W} \, dx$$

as the "<u>residues</u>" of χ_1 and χ_2 respectively at s. Note that this can be dependent on C. Since in a picture like:

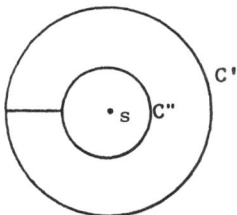

the multivalued nature of the integrand means that the integrals along the short segment need not cancel.

However when the growth conditions of Proposition I.4.4 hold for $\chi_1(x),\ \chi_2(x)$ with $N < 1$ at s we can shrink C to s and get

<u>Proposition I.4.8</u>: $[m_\gamma, n_\gamma] = ([\chi_1(\gamma_1),\ \chi_2(\gamma_1)] + [c_1, c_2])(M_\gamma - I)$ where γ_1 is a path from x_0 to s.

<u>Proof</u>: The residues in the limit are zero by our previous arguments. $\square$

All the above formulas will be useful when we interpret periods in terms of parabolic cohomology and in the specific case of elliptic surfaces. In particular this last proposition is equivalent to one of the requirements in the definition of a parabolic cocycle (see Part II, §3).

Part II.   K-EQUATIONS

§1.   Definitions.

Let X be a complete smooth connected algebraic curve over $\mathbb{C}$ with function field denoted by $K(X)$. After fixing a parameter $x \in K(X)$, consider an algebraic differential equation on X

$$1) \qquad \Lambda f = \frac{d^2 f}{dx^2} + P \frac{df}{dx} + Qf = 0$$

with P and Q in $K(X)$ and f an unknown function.

Definition II.1.1: $\Lambda f = 0$ is called a K-equation if it possesses two solutions, ω_1 and ω_2, which are holomorphic non-vanishing multivalued functions on some Zariski open subset X_0 of X, satisfying

(i) ω_1 and ω_2 form a basis of solutions,

(ii)   for every closed path $\gamma \in \pi_1(X_0)$ the analytic continuation of $\binom{\omega_1}{\omega_2}$ around γ is $M_\gamma \binom{\omega_1}{\omega_2}$ with $M_\gamma \in SL_2(\mathbb{Z})$ (the monodromy representation),

(iii) $\text{Im}(\omega_1/\omega_2) > 0$ on X_0 (positivity).

Such a pair of solutions is called a K-basis. In addition, since the monodromy is in $SL_2(\mathbb{Z})$, the Wronskian $W = e^{-\int P dx}$ is single-valued. We assume as part of our definition:

(iv) $W \in K(X)$.

This last is in fact equivalent to Λ having regular singular points.   □

Let $\Lambda f = 0$ be a K-equation with K-basis ω_1 and ω_2. Consider the function $J = J \circ \omega_1/\omega_2$,

$$X_0 \xrightarrow{\ \omega_1/\omega_2\ } \hbar \xrightarrow{\ J\ } \mathbb{C}$$

where J is the elliptic modular function on the upper-half-plane $\hbar$. This $\hbar$ is a single-valued holomorphic function on $X_0 \subset X$.

Proposition II.1.2: $J \in K(X)$.

Proof: This is an application of a result which appears in Kodaira [16] as Theorem 7.3. It can also be derived as a consequence of Λ having regular singularities. □

We now determine all K-equations. Fix a K-equation $\Lambda f = 0$ on X with K-basis ω_1, ω_2 and let $J = J(\omega_1/\omega_2)$. Say

$$\Lambda f = \frac{d^2 f}{dx^2} + P \frac{df}{dx} + Q f = 0.$$

Theorem II.1.3: There exists an algebraic function λ on X with $\lambda^2 \in K(X)$ such that

$$P = \frac{\left(\frac{dJ}{dx}\right)^2 - J \frac{d^2 J}{dx^2}}{J \frac{dJ}{dx}} - \frac{d}{dx} \log \lambda^2$$

$$Q = \frac{\left(\frac{dJ}{dx}\right)^2 \left(\frac{31}{144} J - \frac{1}{36}\right)}{J^2 (J - 1)^2} - \frac{\left(\frac{dJ}{dx}\right)^2 - J \frac{d^2 J}{dx^2}}{J \frac{dJ}{dx}} \cdot \frac{d}{dx} \log \lambda$$

$$- \frac{\left(\frac{d^2 \lambda}{dx^2}\right)}{\lambda} + 2 \left(\frac{\frac{d\lambda}{dx}}{\lambda}\right)^2.$$

This is known as the K-equation $\Lambda_{(J, \lambda)}$.

Proof: See Stiller [28]. □

It is shown in Stiller [28] that K-equations are precisely those differential equations which arise as the Gauss–Manin connections associated to elliptic surfaces.

Corollary II.1.4: The monodromy group $M \subset SL_2(\mathbb{Z})$ of a K-equation with respect to a fixed K-basis of solutions has finite index in $SL_2(\mathbb{Z})$.

Proof: See Stiller [28]. □

§2. Local properties.

At this point, we calculate the local behaviour of $\Lambda_{(J,\lambda)}$ at a point in $X - X_0$. By Theorem II.1.3

$$\Lambda_{(J,\lambda)}f = \frac{d^2f}{dx^2} + P\frac{df}{dx} + Qf = 0$$

where

$$P = \frac{(\frac{dJ}{dx})^2 - J\frac{d^2J}{dx^2}}{J\frac{dJ}{dx}} - \frac{d}{dx}\log\lambda^2$$

$$Q = \frac{(\frac{dJ}{dx})^2(\frac{31}{144}J - \frac{1}{36})}{J^2(J-1)^2} - \frac{(\frac{dJ}{dx})^2 - J\frac{d^2J}{dx^2}}{J\frac{dJ}{dx}}\frac{d}{dx}\log\lambda$$

$$- \frac{\left(\frac{d^2\lambda}{dx^2}\right)}{\lambda} + 2\left(\frac{\frac{d\lambda}{dx}}{\lambda}\right)^2 .$$

Now the "parameter" x does not provide a good local parameter everywhere and could therefore introduce complications. This however is <u>not</u> the case; if t is a local parameter at any point in X then when we express $\Lambda_{(J,\lambda)}$ locally it retains the above form with t in place of x, i.e.

$$P = \frac{(\frac{dJ}{dt})^2 - J\frac{d^2J}{dt^2}}{J\frac{dJ}{dt}} - \frac{d}{dt}\log\lambda^2$$

$$Q = \text{etc.}$$

We see that the global "parameter" x plays no role in determining the singularities. Thus the above form can be used in local calculations.

We will begin by calculating the local behaviour of the K-equation $\Lambda_{(J,1)}$ (i.e. assume $\lambda = 1$):

Case 1: J has a pole of order $n \geq 1$. In terms of a local coordinate t, $J = \dfrac{c_{-n}}{t^n} + \dots$. An easy calculation gives

$$P = \frac{1}{t} + \text{holomorphic}$$

$$Q = \frac{0}{t^2} + \frac{q_{-1}}{t} + \text{holomorphic}.$$

Since our equation has regular singular points (as we shall see in the course of these calculations) we can apply well known techniques to determine the solutions (see Ince [14], Griffiths [10], or Deligne [5]). The indicial equation is

$$I(v) = v^2 + (p_{-1} - 1)v + q_{-2}$$

where $P = \dfrac{p_{-1}}{t} + \dots$ and $Q = \dfrac{q_{-2}}{t^2} + \dots$. In this case

$$I(v) = v^2$$

and the exponents are $0,0$. Thus the equation has a basis of solutions of the form

$$u_2 = 1 + \dots$$

$$u_1 = u_2 \left(\frac{1}{2\pi i} \log t + \text{holomorphic} \right)$$

so that the local monodromy is $\left(\begin{smallmatrix} 1 & 1 \\ 0 & 1 \end{smallmatrix} \right)$. To reintroduce λ is simple. $\Lambda_{(J,\lambda)}$ will have solutions

$$\lambda u_1, \ \lambda u_2.$$

So as $\lambda^2 \in K(X)$, if $\lambda^2 = t^r \cdot$ (holomorphic non-vanishing) then the new exponents are $\frac{r}{2}, \frac{r}{2}$ and the local monodromy is $\pm\begin{pmatrix} 1 & 1 \\ 0 & 1 \end{pmatrix}$. Note that u_1, u_2 need not be a K-basis. Using Kodaira'a classification for the monodromy at the singular fibers of an elliptic surface, one can show that if ω_1, ω_2 are a K-basis then after "shifting the cusp to ∞" we have that the continuation of ω_1, ω_2 around the singularity gives $\pm\begin{pmatrix} 1 & n \\ 0 & 1 \end{pmatrix}$ where $n > 0$ is the order of the pole of $J = J(\omega_1/\omega_2)$.

Case 2: J has a zero of order $n \geq 1$. Then

$$P = \frac{1}{t} + \text{holomorphic}$$

$$Q = \frac{-\frac{n^2}{36}}{t^2} + \dots \, .$$

The indicial equation is

$$I(v) = v^2 - \frac{n^2}{36}$$

so the exponents are $\pm\frac{n}{6}$. If $n \not\equiv 0$ mod 3 the equation has a basis of solutions of the form

$$u_1 = t^{-n/6}(1 + \text{higher order in } t)$$
$$u_2 = t^{n/6} (1 + \text{higher order in } t).$$

If $n \equiv 0$ mod 3 one must check certain higher order conditions to insure that no logarithmic behaviour occurs. One may also appeal again to the theory of elliptic surfaces where, since $J = 0$, the monodromy must be of finite order. Thus even when $n \equiv 0$ mod 3

$$u_1 = t^{-n/6}(1 + \text{higher order in } t)$$
$$u_2 = t^{n/6} (1 + \text{higher order in } t)$$

still gives a basis for the space of solutions and the monodromy is $\pm\begin{pmatrix} 1 & 0 \\ 0 & 1 \end{pmatrix}$.

Case 3: $J = 1$ to order $n \geq 1$. Then

$$P = \frac{-n + 1}{t} + \text{holomorphic}$$

$$Q = \frac{\frac{3}{16} n^2}{t^2} + \dots .$$

The indical equation is

$$I(v) = v^2 - nv + \frac{3}{16} n^2$$

so the exponents are $n/4$ and $3n/4$. If n is odd then the exponents differ by $\frac{n}{2} \notin \mathbb{Z}$ so there is a basis of solutions of the form

$$u = t^{n/4} (1 + \text{higher order terms in } t)$$

$$u = t^{3n/4} (1 + \text{higher order terms in } t).$$

As before, when n is even we must check certain higher order conditions in order to show that there is no logarithmic behaviour and this can be done. Thus the above form is valid even when n is even.

Case 4: $J \neq 0, 1, \infty$ but $\text{ord } dJ = n \geq 1$. We then have

$$P = \frac{-n}{t} + \dots$$

$$Q = \text{holomorphic}.$$

So the indicial equation is $I(v) = v^2 + (-n-1)v$ and the exponents are $(0, n+1)$. Since these differ by an integer, higher order conditions must again be checked. However all is well, and there is a basis of solutions of the form

$$u_1 = 1 + \text{higher order terms in } t$$

$$u_2 = t^{n+1} (1 + \text{higher order terms in } t).$$

Note that at any singular point where the exponents are distinct integers (a cosingular point) any K-basis ω_1, ω_2 of Λ must extend with ω_1, ω_2 meromorphic; both with order equal to the lesser exponent. This follows because ω_1/ω_2 (or ω_2/ω_1) is holomorphic at $t = 0$ with positive (or negative) imaginary part at all nearby points. Thus ω_1/ω_2 is holomorphic non-vanishing and the period map into $\hbar$ is well defined.

On the other hand

$$\frac{d}{dt}\,(\omega_1/\omega_2) \quad \text{is} \quad \frac{-W}{\omega_2^2}$$

where W is the Wronskian of Λ computed with respect to t, which is $\frac{dJ/dt}{J}$ or more generally $\lambda^2\,\frac{dJ/dt}{J}$ up to a constant multiple. Thus $\frac{d}{dt}(\omega_1/\omega_2)$ vanishes at these points (provided the exponents differ by 2 or more) and the period map from a neighborhood of $t = 0$ to $\hbar$ given by ω_1/ω_2 is ramified. In the case where the exponents differ by 1 the period map is unramified.

Case 5: At all remaining points $\Lambda_{(J,1)}$ is holomorphic, and any K-basis ω_1, ω_2 will be holomorphic and non-vanishing.

Finally, to pass from $\Lambda_{(J,1)}$ to a description of the singularities of $\Lambda_{(J,\lambda)}$ is very simple and is carried out as outlined in Case 1.

§3. Automorphic forms associated to K-equations and parabolic cohomology.

In this section we shall specialize our previous results to the case where the differential equation Λ is a K-equation and in particular we want to elaborate fully the relationship between the periods and parabolic cohomology that was mentioned only briefly above.

Let $\Lambda_{(J,\lambda)}$ be a K-equation on X and S ⊂ X the set of true singular

points, that is the set of points where the local monodromy is not
trivial. From our local calculations in section §2 above, the set of __all__
singular points, true singularities as well as cosingularities, is seen to
be contained in the set of points where $\lambda^2 = 0$, ∞, $J = 0$, 1, ∞, or
$J \neq 0$, 1, ∞ but ord $dJ > 0$ with the parameter $x \in K(X)$ playing no role.

On the Zariski open set $X_0 = X - S$ we can find a "K-basis" ω_1, ω_2
which are __meromorphic__ multivalued functions whose global monodromy group M
is contained as a subgroup of finite index in $SL_2(\mathbb{Z})$ and having
$Im(\omega_1/\omega_2) > 0$ with $J(\omega_1/\omega_2) = J$. We remark that ω_1, ω_2 are unique up
to the natural action of $SL_2(\mathbb{Z})$ and scalar matrices $\begin{pmatrix} a & 0 \\ 0 & a \end{pmatrix} \in GL(\mathbb{C})$.
Notice also that if we remove from X_0 the set of cosingular points, then
on the remaining Zariski open subset ω_1, ω_2 will be holomorphic
non-vanishing multivalued functions with $\omega = \omega_1/\omega_2$ extending to a
holomorphic non-vanishing multivalued function having $Im\ \omega > 0$ on __all__ of
X_0 (see §2 above).

Thus we can define the period map $\omega = \omega_1/\omega_2$ as in Part I, §1:

1)
$$
\begin{array}{ccc}
\mathcal{h},\ z_0 & \xrightarrow{\ \omega\ } & \mathcal{h},\ \omega(z_0) \\
\pi \downarrow & & \\
X_0,\ x_0 & &
\end{array}
$$

where $x_0 \in X_0$ is a base point selected away from the cosingular points
for convenience. Notice that S always contains at least one point
because J is non-constant and any pole of J will be a true
singularity. Moreover, it is easy to show that if $X \cong P_{\mathbb{C}}^1$ then S must
contain at least 3 points. Thus the universal cover of X_0 will be
isomorphic to $\mathcal{h}$. The period mapping will be ramified at exactly the
subset of cosingular points where the exponent difference is 2 or more (as
opposed to 1 which is the only other possibility) and the order of

ramification will be that difference (see remarks at the end of §2 above).

If we let $\overline{M}$ denote the global projective monodromy group in $\text{PSL}_2(\mathbb{Z})$,

$$\overline{M} = M \cdot \{\pm I\}/\{\pm I\} \subset \text{PSL}_2(\mathbb{Z}),$$

then we can complete diagram 1) to give:

where C_0 is the modular curve $\hbar/\overline{M}$. We denote the compactification of C_0 by C, and it is clear that the induced map $X_0 \xrightarrow{[\omega]} C_0$ extends to a map $X \xrightarrow{[\omega]} C$ by virtue of the regularity of Λ. Moreover, if we consider the diagram

where the map from $C_0 = \hbar/\overline{M} \to \mathbb{C} = \hbar/\text{PSL}_2(\mathbb{Z})$ is the natural one, then the resulting composite map

$$X_0 \xrightarrow{[\omega]} C_0 \longrightarrow \mathbb{C}$$

when extended

$$X \xrightarrow{[\omega]} C \longrightarrow P^1_\mathbb{C}$$

is just the rational function $J \in K(X)$. This yields:

<u>Proposition II.3.1</u>: The index of the global projective monodromy group $\overline{M}$ in $PSL_2(\mathbb{Z})$ divides the valence of J. (The valence of a rational function is just its total number of poles counted with multiplicity.) □

Notice that if we change the basis ω_1, ω_2 for the space of solutions to our K-equation $\Lambda_{(J,\lambda)}$ by a matrix in $SL_2(\mathbb{Z})$ and/or a scalar matrix $\begin{pmatrix} a & 0 \\ 0 & a \end{pmatrix} \in GL_2(\mathbb{C})$ the result will be a new K-basis $\tilde{\omega}_1$, $\tilde{\omega}_2$ with $J = J(\tilde{\omega}_1/\tilde{\omega}_2) = J(\omega_1/\omega_2)$ being preserved. The effect of such a change is to conjugate the monodromy representation and the global monodromy group M (or $\overline{M}$) by an element of $SL_2(\mathbb{Z})$ (or $PSL_2(\mathbb{Z})$) and this will give another modular curve $\tilde{C}_0$ isomorphic to C_0 and via this isomorphism the diagram

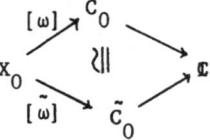

commutes. We remark that it is possible for a K-equation Λ to have two distinct K-bases ω_1, ω_2 and $\tilde{\omega}_1$, $\tilde{\omega}_2$ with

$$J(\tilde{\omega}_1/\tilde{\omega}_2) = \tilde{J} \neq J = J(\omega_1/\omega_2),$$

in other words one can have

$$\Lambda_{(J,\lambda)} = \Lambda_{(\tilde{J},\tilde{\lambda})}$$

with $J \neq \tilde{J}$. In this case it can be shown that the change of basis between ω_1, ω_2 and $\tilde{\omega}_1$, $\tilde{\omega}_2$ is given by a matrix in $GL_2^+(\mathbb{Q})$. The corresponding modular curves will still be isomorphic as above and the index and valence will be preserved. This situation is related to an isogeny phenomenon in the theory of elliptic surfaces (see Stiller [28] and [30]), and along with

Proposition II.3.1 is useful in analysis of the torsion part of the

Mordell-Weil group of the corresponding elliptic curve defined over the

function field of X (see Cox [3] and remarks in Part IV).

We would now like to turn out interest back to the inhomogeneous

equations

$$\Lambda f = Z$$

where $\Lambda = \Lambda_{(J,\lambda)}$ is a K-equation and $Z \in K(X)$. From Theorem I.2.1 we

have

$$f = (\int_{x_0}^{x} \frac{-\omega_2 Z}{W} \, dx + c_1)\omega_1 + (\int_{x_0}^{x} \frac{\omega_1 X}{W} \, dx + c_2)\omega_2$$

where for the moment we will assume that Z is regular at x_0 and that

$x \in K(X)$ is a good local parameter at x_0. The above solution then makes

sense on some Zariski open subset of X_0 as a multivalued holomorphic

function. Via the integral transformations of Part I, §3 we can associate

to f an "automorphic form" g, which by Theorem I.3.1 is

$$g = \frac{\omega_2^3 Z}{W^2}.$$

Recall that on $\hbar$ we have

$$g(\gamma z) = (c_\gamma \omega(z) + d_\gamma)^3 \, g(z)$$

where $z \in \hbar$, $\gamma \in \overline{\Gamma} \subset PSL_2(\mathbb{R})$, and $\begin{pmatrix} a_\gamma & b_\gamma \\ c_\gamma & d_\gamma \end{pmatrix}$ is the monodromy as in

Part I, §3.

We would like to examine in what sense g is an automorphic form.

Suppose for the moment that the global monodromy group $M \subset SL_2(\mathbb{Z})$ is such

that there exist classical automorphic forms of weight three for M. (So

in particular the matrix $-I_2 = \begin{pmatrix} -1 & 0 \\ 0 & -1 \end{pmatrix}$ is not in M.) Let h be any

(possibly meromorphic). Then pulling back via the period map gives a

function $h \circ \omega$ on $\hbar$ which transforms as

(2) $\qquad (h \circ \omega)(\gamma z) = (c_\gamma \omega(z) + d_\gamma)^3 (h \circ \omega)(z) \qquad \gamma \in \bar{\Gamma}.$

Thus these pull-backs transform like the automorphic forms associated to the inhomogeneous equations $\Lambda_{(J,\lambda)} f = Z$, $Z \in K(X)$. This shows that we may regard the automorphic forms attached to inhomogeneous equations formed with a K-equation as meromorphic sections of the pull-back, via the map induced by the period map, of the line bundle associated to the classical automorphic forms of weight 3 for the modular curve $\hbar/\overline{M}$ (see Bayer and Neukirch [2]). This will be useful later on in Parts III and IV.

Jumping ahead a little and introducing some elliptic surfaces (the reader can consult Part III for details), suppose as above that M does not contain $-I_2$, then using a construction of Shioda [27] one can build a canonical basic elliptic surface over $C = \hbar^*/\overline{M}$ whose functional invariant is given by the natural map

$$C = \hbar^*/\overline{M} \to \hbar^*/PSL_2(\mathbb{Z}) = \mathbb{P}^1_{\mathfrak{C}}.$$

(In other words it is the elliptic modular function J viewed as a modular function for M.) It is an easy exercise to verify that the pull-back of this surface to one over X via $[\omega]$ induces the one associated to $\Lambda_{(J,\lambda)}$ on X.

An interesting case of all we have been discussing is the case where ω is the identity map. This means that up to isomorphism we are dealing with one of the elliptic modular surfaces of Shioda as described above. There the forms that arise from the inhomogeneous equations are the usual forms of weight three. Also there is an interesting connection between forms of weight one for the group M and solutions to the differential equation for the elliptic modular surface (see Stiller [31]).

To close this section we want to consider the periods and their relation to parabolic cohomology. Let $\Lambda = \Lambda_{(J,\lambda)}$ be a K-equation on X

with K-basis ω_1, ω_2. As above we let X_0 denote the Zariski open subset of X obtained by removing the set S of true singular points of Λ.

Definition II.3.2: A rational function $Z \in K(X)$ will be said to satisfy the residue condition for Λ if the differential forms

$$\frac{-\omega_2 Z}{W} dx \qquad \frac{\omega_1 Z}{W} dx$$

have zero residue at every point in X_0. Note that this condition is completely independent of the choice of basis ω_1, ω_2. We denote by L_Λ^{res} the set of $Z \in K(X)$ which satisfy the residue condition for Λ.   □

We now consider an inhomogeneous equation $\Lambda f = Z$ with $Z \in L_\Lambda^{res}$. Then as we have seen

$$f = (\int_{x_0}^{x} \frac{-\omega_2 Z}{W} dx + c_1)\omega_1 + (\int_{x_0}^{x} \frac{\omega_1 Z}{W} dx + c_2)\omega_2$$

and the associated automorphic form is

$$g = \frac{\omega_2^3 Z}{W^2} .$$

As before for $\gamma \in \overline{\Gamma} \subset PSL_2(\mathbb{R})$ corresponding to $\gamma \in \pi_1(X_0,x_0)$ via the uniformization $\pi : \mathcal{h}$, $z_0 \rightarrow X_0$, x_0, we have

$$f \rightarrow f + m_\gamma \omega_1 + n_\gamma \omega_2 \qquad m_\gamma, h_\gamma \in \mathbb{C}$$

where m_γ, n_γ are the periods. Notice that the periods are well-defined independent of the specific path chosen to represent the class $\gamma \in \pi_1(X_0,x_0)$ because $Z \in L_\Lambda^{res}$.

Recall the transformation law given in Proposition I.4.5, namely:

$$[m_{\gamma\tau}, n_{\gamma\tau}] = [m_\gamma, n_\gamma]M_\tau + [m_\tau, n_\tau]$$

where $\gamma\tau$ is the path gotten by following first γ and then τ. This transformation law is one of the conditions for giving a parabolic cocycle with respect to a given representation of $\overline{\Gamma}$ (see Shimura [26]; note that our definition is slightly different from that of Shimura). There is a second condition, namely for every parabolic element $\gamma \in \overline{\Gamma}$ there must exist a vector $[a_1, a_2]$ with

(4) $$[m_\gamma, n_\gamma] = [a_1, a_2](M_\gamma - I).$$

For example the growth condition in Proposition I.4.8 is sufficient for this. In a moment we shall state necessary and sufficient conditions.

Recall that Proposition I.4.2 gives

$$[m_\gamma, n_\gamma] = [\chi_1(\gamma), \chi_2(\gamma)]M_\gamma + [c_1, c_2](M_\gamma - I).$$

The term $[\chi_1(\gamma), \chi_2(\gamma)]M_\gamma$ does not depend on c_1, c_2 and the term $[c_1, c_2](M_\gamma - I)$ is a parabolic coboundary. Thus condition (4) above is independent of the choice of a particular solution of the inhomogeneous equation

$$\Lambda f = Z.$$

Moreover assuming condition (4) the resulting parabolic cohomology class depends only on the inhomogeneous equation not on the choice of a particular solution.

Now suppose we choose another K-basis, say

$$\begin{pmatrix} \tilde{\omega}_1 \\ \tilde{\omega}_2 \end{pmatrix} = M \begin{pmatrix} \omega_1 \\ \omega_2 \end{pmatrix} \qquad M \in SL_2(\mathbb{Z}).$$

Then with respect to this new basis we can write our original solution

$$f = \left(\int_{x_0}^{x} \frac{-\omega_2 Z}{W} \, dx + c_1 \right) \omega_1 + \left(\int_{x_0}^{x} \frac{\omega_1 Z}{W} \, dx + c_2 \right) \omega_2$$

as

$$f = \left(\int_{x_0}^{x} \frac{-\tilde{\omega}_2 Z}{W} \, dx + \tilde{c}_1 \right) \tilde{\omega}_1 + \left(\int_{x_0}^{x} \frac{\tilde{\omega}_1 Z}{W} \, dx + \tilde{c}_2 \right) \tilde{\omega}_2$$

where

$$[\tilde{c}_1, \tilde{c}_2] = [c_1, c_2] M^{-1}.$$

In addition,

$$[\tilde{\chi}_1(\gamma), \ \tilde{\chi}_2(\gamma)] = [\chi_1(\gamma), \ \chi_2(\gamma)] M^{-1}$$

and

$$[\tilde{m}_\gamma, \ \tilde{n}_\gamma] = [m_\gamma, \ n_\gamma] M^{-1}.$$

Of course changing the basis changes the monodromy representation by conjugation, and gives a new representation of $\bar{\Gamma}$; namely

$$\bar{\Gamma} \to SL_2(\mathbb{Z})$$

$$\gamma \to M M_\gamma M^{-1} = \tilde{M}_\gamma.$$

Condition (4) above for the parabolic cochain $[\tilde{m}_\gamma, \ \tilde{n}_\gamma]$ with respect to the new representation is that for every parabolic element $\gamma \in \bar{\Gamma}$ there exists $[\tilde{a}_1, \tilde{a}_2]$ such that

$$[\tilde{m}_\gamma, \ \tilde{n}_\gamma] = [\tilde{a}_1, \ \tilde{a}_2](\tilde{M}_\gamma - I),$$

but this is equivalent to $[\tilde{m}_\gamma, \ \tilde{n}_\gamma] = [\tilde{a}_1, \ \tilde{a}_2] M(M_\gamma - I) M^{-1}$ or, multiplying both sides by M on the right,

$$[m_\gamma, \ n_\gamma] = [a_1, \ a_2](M_\gamma - I)^{-1}$$

where $[a_1, a_2] = [\bar{a}_1, \bar{a}_2]M$. Thus condition (4) is independent of the basis chosen -- if it holds for one basis, it holds for any other.

Let $\gamma \in \bar{\Gamma}$ be a parabolic element, then the fixed point of γ is a cusp for $\bar{\Gamma}$ acting on $\mathcal{h}$ which lies over some point $s \in S$. Now γ also corresponds to a path $\gamma \in \pi_1(X_0, x_0)$ which without loss of generality can be taken to be a multiple of a simple loop around s, i.e.

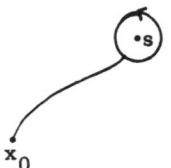

It is clearly enough to consider just the case of a simple loop. Condition (4) is immediate except in the case where the local monodromy around s is parabolic with trace 2. For a K-equation $\Lambda = \Lambda_{(J, \lambda)}$ this occurs at the points where J has a pole and λ^2 has even order -- the exponents will be (r, r) $r \in \mathbb{Z}$ (see §2 above). Notice that at such an s the equation Λ will have a one-dimensional space of invariant solutions, i.e. single-valued solutions. Choose a K-basis ω_1, ω_2 at x_0 so that the monodromy around the simple loop γ is $\begin{pmatrix} 1 & n \\ 0 & 1 \end{pmatrix}$ $n \in \mathbb{Z} - \{0\}$. Notice that ω_2 is then an invariant solution near s. This choice of basis is equivalent to choosing ω_1, ω_2 so that the period map $\omega: \mathcal{h} \to \mathcal{h}$ takes the cusp fixed by γ to the cusp at infinity.

Condition (4) is now

$$[m_\gamma, n_\gamma] = [a_1, a_2]\left(\begin{pmatrix} 1 & n \\ 0 & 1 \end{pmatrix} - I\right)$$

which can be satisfied if and only if

$$m_\gamma = 0$$

and this in turn is equivalent to

$$\chi_1(\gamma) = 0 .$$

Thus $\int_\gamma \dfrac{-\omega_2 Z}{W} \, dx$ must be zero which means that the differential $\dfrac{-\omega_2 Z}{W} \, dx$ must have zero residue at s. This leads us to:

<u>Definition II.3.3</u>: A rational function $Z \in K(X)$ will be said to satisfy the **parabolic residue conditions** for Λ if the differential forms

$$\frac{-\omega_2 Z}{W} \, dx \qquad\qquad \frac{\omega_1 Z}{W} \, dx$$

have zero residue at every point in X_0 and if at every point $s \in S$ where the monodromy is conjugate to $\begin{pmatrix} 1 & n \\ 0 & 1 \end{pmatrix}$ $n \in \mathbb{Z} - \{0\}$, the differential form

$$\frac{\tilde{\omega} Z}{W} \, dx$$

has zero residue where $\tilde{\omega}$ is a local invariant solution at s. We denote by L_Λ^{para} the subset of $Z \in K(X)$ which satisfies the parabolic residue conditions for Λ. Note $L_\Lambda^{para} \subset L_\Lambda^{res}$. $\qquad\qquad \square$

This condition will be important when we deal with sections of an elliptic surface in Part III.

Thus:

<u>Theorem II.3.4</u>: Let Λ be a K-equation with K-basis ω_1, ω_2. For every $Z \in L_\Lambda^{para}$ the periods of the inhomogeneous equation $\Lambda f = Z$ with respect to ω_1, ω_2 give rise to a parabolic cohomology class with respect to the monodromy representation. This class depends only on Z, not on the choice of a particular solution f. Moreover Z gives rise to the zero class if and only if $Z = \Lambda Z'$ for some $Z' \in K(X)$. Thus the sequence

$$0 \rightarrow L_\Lambda^{para} \cap \{\Lambda Z': Z' \in K(X)\} \rightarrow L_\Lambda^{para} \rightarrow H^{para}$$

is exact where H^{para} is the parabolic cohomology group.

Proof: All except the last part appears above. The last part is Theorem
I.4.3.   □

Much of this discussion of course generalizes to arbitrary Λ. In a
sense the conditions defined above are a rough analogue of differentials of
the second kind (as we shall see later). One can also follow the analogue
for classical automorphic forms as in Shimura [26] defining an inner
product for the forms associated to appropriate inhomogeneous equations
(i.e. those with our restrictions above) as Hoyt [12] does for sections of
an elliptic surface and the forms associated to them.

It can also be shown that the last map in Theorem II.3.4 from L_Λ^{para}
to H^{para} is in fact onto. (For some interpretations of this isomorphism
·and its relationship to the classical Shimura isomorphism in the theory of
automorphic forms see Part IV and Stiller [36].)

PART III.   ELLIPTIC SURFACES

§1.   Introduction.

Let E be an elliptic surface having a global section over its base
curve X. We assume throughout that the functional invariant J is
non-constant and that E has no exceptional curves of the first kind in
the fibers. We shall frequently refer to such an elliptic surface as a
basic elliptic surface. The generic fiber E^{gen} is naturally an elliptic
curve over the function field $K(X)$ of the base curve with the section
providing a $K(X)$-rational point to serve as zero in the group. We shall be
interested in the group of $K(X)$-rational points on E^{gen}, denoted
$E^{gen}(K(X))$, which corresponds to the group of sections of E over X.
This is a finitely generated abelian group whose rank we denote by r.

Let $[\mathcal{O}\!L]$ be any $K(X)$-rational divisor class of degree zero on E^{gen}.
Choose a model of E^{gen}:

$$Y^2 = 4X^3 - g_2 X - g_3 \qquad g_2,\ g_3 \ \varepsilon \ K(X).$$

In Manin [20] a homomorphism is constructed which maps divisor classes of
degree zero rational over $K(X)$ to the additive group $K(X)$. Namely, one
takes the Poincaré normal function

$$f = \int^{\mathcal{O}\!L} \frac{dX}{Y} = \sum_{\mathcal{P}} \nu_{\mathcal{P}}(\mathcal{O}\!L) \int_{\mathcal{P}_0}^{\mathcal{P}} \frac{dX}{Y} \ ,$$

where $\mathcal{P}_0$ is arbitrary and $\nu_{\mathcal{P}}(\mathcal{O}\!L)$ is the order of $\mathcal{O}\!L$ at $\mathcal{P}$, and then
applies to it the operator Λ annihilating the periods of $\frac{dX}{Y}$. Since f
is determined only up to periods the result is a single-valued and in fact
rational function Z in $K(X)$ depending on the divisor class of $\mathcal{O}\!L$.

In this way we can view f as the solution of an inhomogeneous
equation

$$\Lambda f = Z,$$

with Z depending on the choice of model for E^{gen} over K(X) and on the
choice of the parameter $x \in K(X)$ used in forming Λ.

Because E^{gen} is an elliptic curve, any K(X)-rational divisor α of
degree zero is linearly equivalent to a divisor $\wp - \theta$ where
$\wp \in E^{gen}(K(X))$ and θ is the zero element. Therefore the above map can
be realized as a homomorphism:

$$E^{gen}(K(X)) \longrightarrow K(X)$$

$$\wp \longrightarrow \Lambda \int_{\theta}^{\wp} \frac{dX}{Y} = Z.$$

In this way we associate to each K(X)-rational point on E^{gen}, or what is
the same to each section of E over X, a rational function Z in K(X).
For further details see Stiller [29].

From another point of view, choose a Zariski-open subset $X_0 \subset X$ on
which the operator Λ annihilating the periods of $\frac{dX}{Y}$ for the chosen
model of the generic fiber is holomorphic. Say

$$\Lambda = \frac{d^2}{dx^2} + P \frac{d}{dx} + Q$$

for some parameter $x \in K(X)$ which for the moment we will assume is a good
local parameter on all of X_0. We also fix a base point $x_0 \in X_0$. Note
that the bad fibers of E/X occur at exactly the set of <u>true</u> singularities
of Λ (Stiller [28]), so that over X_0 all the fibers are elliptic
curves.

Given a section s of E over X we can integrate the

specialization of $\frac{dX}{Y}$ to a differential of the first kind on each fiber near x_0 along a path between the zero section $\mathcal{O}$ and s in the fiber. This makes sense because near a good fiber the family is topologically trivial. The result is an analytic function f defined in a neighborhood of x_0 which is unique up to addition of $m\omega_1 + n\omega_2$ where $m, n \in \mathbb{Z}$ and ω_1, ω_2 are functions which give a basis for the periods in each fiber. This function f admits unrestricted analytic continuation throughout X_0 and gives a multivalued holomorphic function on X_0 which transforms around $\gamma \in \pi_1(X_0, x_0)$ by

$$f \to f + m_\gamma \omega_1 + n_\gamma \omega_2 \qquad m_\gamma, n_\gamma \in \mathbb{Z} .$$

Applying Λ to f yields a single-valued, in fact rational function

$$\Lambda f = Z,$$

the same rational function mentioned above.

Much of the above depends on a choice of model for E^{gen} over $K(X)$ and in what follows we shall frequently need to change the model in order to obtain our results. For this reason, we shall take some time here to explain how the various objects we have defined change when the model is changed.

It is well-known that we can find a function λ (unique up to sign) with $\lambda^2 \in K(X)$ so that

$$\lambda^{-4} \frac{27J}{J-1} = g_2 \quad \text{and} \quad \lambda^{-6} \frac{27J}{J-1} = g_3$$

gives the model for E^{gen} over $K(X)$ as above.

__Theorem III.1.1__: The K-equation $\Lambda_{(J, \lambda)}$ annihilates the periods of $\frac{dX}{Y}$ for the curve with $g_2 = \lambda^{-4} \frac{27J}{J-1}$ and $g_3 = \lambda^{-6} \frac{27J}{J-1}$.

Proof: See Stiller [29].   □

An associated K-basis of this equation (see Part II, §1 above) ω_1, ω_2

then provides a basis for the solutions of the homogeneous equation

$$\Lambda_{(J,\lambda)} f = 0$$

and represents the periods of $\frac{dX}{Y}$ as we move from fiber to fiber.

In passing to an isomorphic model over $K(X)$, or equivalently using

another $K(X)$-rational differential of the first kind, the periods are

simply multiplied by a function $g \in K(X)$. The new model has

$$g_2' = g^{-4} \lambda^{-4} \frac{27J}{J-1} \quad \text{and} \quad g_3' = g^{-6} \lambda^{-6} \frac{27J}{J-1}$$

i.e.

$$\tilde{Y}^2 = 4\tilde{X}^3 - g_2'\tilde{X} - g_3' \quad .$$

The operator which annihilates the periods of $\frac{d\tilde{X}}{\tilde{Y}}$ for this model is

$\Lambda_{(J,\lambda g)}$. The so called "twist" of $\Lambda_{(J,\lambda)}$ by g.

Note that $\Lambda_{(J,\lambda g)}$ has $g\omega_1$, $g\omega_2$ as K-basis. Furthermore if f is

defined as above, i.e. as $\int_\sigma^s \frac{dX}{Y}$ for some section s of E over X, then

in the new model f becomes gf because, as is clear, we have scaled each

lattice by a factor of homothety equal to the value of g. Also

$$\Lambda_{(J,\lambda g)}(gf) = gZ$$

so Z is multiplied by g. Note too that the new Wronskian is g^2 times

the old one. (This is just our earlier notion of the "twist" of an

inhomogeneous equation; see Part I, §2.)

As above, let $f = \int_\sigma^s \frac{dX}{Y}$ for some section s of E over X and

choose $X_0 \subset X$ and a base point $x_0 \in X_0$ as above. Then since f

satisfies $\Lambda f = Z$, Theorem I.2.1 implies that:

Proposition III.1.2: $f = (\int_{x_0}^{x} \frac{-\omega_2 Z}{W} + c_1)\omega_1 + (\int_{x_0}^{x} \frac{\omega_1 Z}{W} + c_2)\omega_2$ where ω_1, ω_2 are a fixed K-basis in a neighborhood of x_0 and c_1, c_2 are constants which reflect that s cuts the fiber E_{x_0} over x_0 at $c_1\omega_1(x_0) + c_2\omega_2(x_0)$ viewing E_{x_0} as $\mathbb{C}/\mathbb{Z}\omega_1(x_0) + \mathbb{Z}\omega_2(x_0)$, and $W = \omega_1\omega_2' - \omega_2\omega_1'$. □

§2. A bound on the rank r of $E^{gen}(K(X))$.

In this section we will construct a bound on the rank r of $E^{gen}(K(X))$ for our given basic elliptic surface E over the base curve X. Recall the homomorphism described in the last section:

$$E^{gen}(K(X)) \longrightarrow K(X)$$

$$\mathscr{P} \longrightarrow \Lambda \int_{\theta}^{\mathscr{P}} \frac{dX}{Y} = Z$$

where Λ is the second order equation on X annihilating the periods of $\frac{dX}{Y}$. As mentioned Λ is a K-equation easily computed once a model for the generic fiber and a parameter are chosen. We would like to give some estimate for the poles of Z so as to obtain via Riemann-Roch a bound for r. This has essentially been done in Stiller [29]. However, there was an error in that paper (the $(dx)^2$ was missing). The correct estimates appear below.

Let E/X be a basic elliptic surface with $E^{gen}/K(X)$ its generic fiber. We denote by θ the section of E/X as well as the origin of $E^{gen}/K(X)$ to which it corresponds. $E^{gen}(K(X))$ will be the group of K(X)-rational points on $E^{gen}/K(X)$. Let B consist of all the points where the projection $E \to X$ is submersive. As is well known (Kodaira [16]), $B \to X$ exhibits B as an analytic family of complex Lie groups over X (or algebraic family of algebraic groups) with fibers

$$B_x = \begin{cases} \text{elliptic curve } E_x \\ \mathbb{C} \times \text{finite group} \\ \mathbb{C}^* \times \text{finite group} \end{cases}$$

depending on the nature of the singular fiber. We let B_0/X be the family obtained by taking the connected component of the identity in each fiber. Let $\Omega(B)$ (resp. $\Omega(B_0)$) be the sheaf of germs of sections of B (resp. B_0). Note that any section of E/X has image in B. Obviously $H^0(X, \Omega(B))$ is in 1-1 correspondence with $E^{gen}(K(X))$. Likewise $H^0(X, \Omega(B_0))$ corresponds to a torsion free subgroup $E_0^{gen}(K(X))$ of finite index (Shioda [27]). For all of our discussion in this section there will be no loss of generality in working with $E_0^{gen}(K(X))$ or any other subgroup of $E^{gen}(K(X))$ of finite index, and we will frequently do so.

We denote by f the line bundle on X obtained by taking the tangent space at σ to each fiber. As is known (Kodaira [17]) the Chern class, $c(f)$, of f equals $-\chi(\mathcal{O}_E)$ the Euler-Poincare characteristic of the structure sheaf $\mathcal{O}_E$. Since J is always assumed non-constant $c(f) \leq -1$.

Let G be the homological invariant. Recall that $G \cong R^1\pi_*\mathbb{Z}$ where $\pi: E \to X$ is the projection and that the stalk G_x of G at $x \in X$ is

$$G_x = \begin{cases} \mathbb{Z} + \mathbb{Z} & \text{good fiber} \\ \mathbb{Z} & \text{type } I_b \quad b > 0 \\ 0 & \text{otherwise.} \end{cases}$$

We then have an exact sequence of sheaves of abelian groups

$$0 \to G \to \mathcal{O}(f) \to \Omega(B_0) \to 0.$$

Now fix a Weierstrass-model of E^{gen},

1) $$Y^2 = 4X^3 - g_2 X - g_3 \qquad g_2, g_3 \in K(X),$$

and let Λ be the K-equation which annihilates the periods ω_1, ω_2 of $\frac{dX}{Y}$ (see §1 above or Stiller [29]). Consider Manin's map

$$E^{gen}(K(X)) \longrightarrow K(X)$$

$$\mathscr{P} \longrightarrow \Lambda \int_{\theta}^{\mathscr{P}} \frac{dX}{Y} = Z(\mathscr{P}).$$

<u>Lemma III.2.1</u>: The kernel of this map if exactly the torsion in $E^{gen}(K(X))$.

<u>Proof</u>: (Manin [21]). □

Thus the image is isomorphic to $\mathbb{Z}^r$ where r is the rank of the free part of $E^{gen}(K(X))$. Note that the image of $E_0^{gen}(K(X))$ in $K(X)$ is also isomorphic to $\mathbb{Z}^r$ and both images have the same $\mathbb{C}$-span in $K(X)$. The same is true for any subgroup of $E^{gen}(K(X))$ of finite index.

At this point we must develop a number of technical results which will aid in the description of the image of this map. Let $\mathscr{P}$ be any point in $E_0^{gen}(K(X))$ viewed as a section of B_0/X. Again denote by f the function $\int_{\theta}^{\mathscr{P}} \frac{dX}{Y}$ which of course depends on the choice of model 1) but not on the parameter chosen as can be seen in Part I, §2 from the remarks following Theorem I.2.1 above. The function f can be viewed as holomorphic multivalued on the Zariski open subset X_0 of X where Λ is holomorphic. If $x \in K(X)$ is the parameter chosen, so that

$$\Lambda = \frac{d^2}{dx^2} + P \frac{d}{dx} + Q,$$

then for every point $x \in X_0$ we can select a local parameter t and one can see that $Z(\frac{dx}{dt})^2$ will be holomorphic. The reason for the factor $(\frac{dx}{dt})^2$ is that when we express Λ in terms of the local parameter t (in which it will have holomorphic coefficients) and then apply it to f, which is invariant of the parameter as remarked, we alter Z by the given factor (see Part I, §2 remarks following Theorem I.2.1). Thus the quadratic differential $Z(dx)^2$ will be holomorphic on X_0.

Recall the exact sequence

$$0 \to G \to \mathcal{O}(\mathcal{f}) \to \Omega(B_0) \to 0.$$

Now there is a natural map, the exponential on fibers

which induces the map $\mathcal{O}(\mathcal{f})$ to $\Omega(B_0)$ mentioned above, and we can interpret $e^{-1}(\mathcal{p})$ as a "multivalued" section of $\mathcal{f}$. If we pick $x_0 \in X_0$ and trivialize $\mathcal{f}$ in a neighborhood U of x_0 in such a way that the sections of G over U correspond to $m\omega_1(u) + n\omega_2(u)$, m, n $\in \mathbb{Z}$, u $\in$ U, then $e^{-1}(\mathcal{p})$ over U corresponds to $f(u) + m\omega_1(u) + n\omega_2(u)$ where $f(u)$ is a branch of $\int_{\theta}^{\mathcal{p}} \frac{dX}{Y}$. We caution that ω_1, ω_2 can degenerate at a point where the fiber does not. Ths is related to the existence of the cosingular points and illustrates the need to exclude them from X_0 for the moment. Later we shall reintroduce them into X_0 as we did in Part II. §3.

Lemma III.2.2: Suppose for some point $\mathcal{p} \in E_0^{gen}(K(X))$ the function $f = \int_{\theta}^{\mathcal{p}} \frac{dX}{Y}$ has a branch which is single-valued on X_0. Then there is a section σ of $\mathcal{f}$ over all of X_0 which lies in $e^{-1}(\mathcal{p})$.

Proof: For suitable trivializations of $\mathcal{f}$ the branches of f can be viewed as giving $e^{-1}(\mathcal{p})$. The result is then clear. □

Our aim now is to show that the function $f = \int_{\theta}^{\mathcal{p}} \frac{dX}{Y}$ can have a single-valued branch on X_0 if and ony if $\mathcal{P} = 0$. Let $i: X_0 \to X$ denote the inclusion map and recall the exact sequence

$$0 \to G \to \mathcal{O}(\mathcal{f}) \xrightarrow{\ e\ } \Omega(B_0) \to 0$$

of sheaves of abelian groups on X. This leads to the commutative diagram

$$\to H^0(X, \mathcal{O}(\mathcal{f})) = 0 \xrightarrow{\ e\ } H^0(X, \Omega(B_0)) \xrightarrow{\ \delta_1\ } H^1(X,G) \to$$

$$\downarrow \qquad\qquad \downarrow r_1 \qquad\qquad \downarrow r_2$$

$$\to H^0(X_0, \mathcal{O}(\mathcal{f})\big|_{X_0}) \xrightarrow{\ e\ } H^0(X_0, \Omega(B_0)\big|_{X_0}) \xrightarrow{\ \delta_2\ } H^1(X_0, G\big|_{X_0}) \to .$$

As remarked earlier the Chern class, $c(\mathcal{f})$, of $\mathcal{f}$ is negative so the line bundle $\mathcal{O}(\mathcal{f})$ has no global sections, i.e. $H^0(X, \mathcal{O}(\mathcal{f})) = 0$.

<u>Theorem III.2.3</u>: For $\mathcal{P} \in E_0^{\text{gen}}(K(X))$ the function $f = \int_\theta^{\mathcal{P}} \frac{dX}{Y}$ can have a branch which is single-valued on X_0 if and only if $\mathcal{P} = 0$.

<u>Proof</u>: By Lemma III.2.2 such a single-valued branch corresponds to a section $\sigma \in H^0(X_0, \mathcal{O}(\mathcal{f})\big|_{X_0})$ whose image under the map e in $H^0(X_0, \Omega(B_0)\big|_{X_0})$ is equal to $r_1(\mathcal{P})$. (Here we have identified $E_0^{\text{gen}}(K(X))$ with $H^0(X, \Omega(B_0))$ as in Kodaira [17] and Shioda [27].)

Since then $0 = \delta_2 e(\sigma) = \delta_2 r_1(\mathcal{P}) = r_2 \delta_1(\mathcal{P})$ and δ_1 is injective, it will be enough to show that the map r_2 is injective to force $\mathcal{P} = 0$.

Consider the Leray spectral sequence for the inclusion map $i: X_0 \to X$ and the sheaf $G\big|_{X_0}$ on X_0. We have

$$E_2^{p,q} = H^p(X, R^q i_* G\big|_{X_0}) \Rightarrow H^*(X_0, G\big|_{X_0})$$

and considering the exact sequence of low order terms we get

$$0 \to H^1(X, i_* G|_{X_0}) \to H^1(X_0, G|_{X_0}) \to H^0(X, R^1 i_* G|_{X_0}) \to$$

Since $i_*(G|_{X_0}) = G$ it is clear that this first map

$$0 \to H^1(X, i_* G|_{X_0}) \to H^1(X_0, G|_{X_0})$$

is r_2. □

<u>Corollary III.2.4</u>: The function $f = \int_0^{\mathcal{P}} \frac{dX}{Y}$ viewed as a solution of the inhomogeneous equation

$$\Lambda f = Z$$

never has trivial periods in the sense of Part I, §4, unless $\mathcal{P}$ is torsion so that $Z = 0$.

<u>Proof</u>: We have

1) $f = (\int_{x_0}^{x} \frac{-\omega_2 Z}{W} dx + c_1)\omega_1 + (\int_{x_0}^{x} \frac{\omega_1 Z}{W} dx + c_2)\omega_2 + m\omega_1 + n\omega_2$

as in Proposition III.1.2 with $m, n \in \mathbb{Z}$. Trivial periods means that there exists $(c_1', c_2') \in \mathbb{C}^2$ such that for every path $\gamma \in \pi_1(X_0, x_0)$ we have the periods $(m_\gamma, n_\gamma) = (c_1', c_2')(M_\gamma - I)$ where M_γ is the monodromy matrix. Now the branches of f in 1) above have integral periods, i.e.

$(m_\gamma, n_\gamma) \in \mathbb{Z}^2$ for all γ (see §1 above). Because the global monodromy group has finite index in $SL_2(\mathbb{Z})$ by Corollary II.1.4, we see that (c_1', c_2') must be in $\mathbb{Q}$. Now $f - c_1'\omega_1 - c_2'\omega_2$ has zero periods and as in Theorem I.4.3 we see that $f - c_1'\omega_1 - c_2'\omega_2 = Z' \in K(X)$. It follows that for a suitable integer N we can arrange to have

$$\Lambda \int_{\mathcal{O}}^{N\wp} \frac{dX}{Y} = \Lambda(Nf) = NZ$$

with $N\wp$ in $E_0^{gen}(K(X))$ and $(c_1', c_2') \in \mathbb{Z}^2$. This implies that $\int_{\mathcal{O}}^{N\wp} \frac{dX}{Y}$ has a single valued branch contrary to the theorem, unless $N\wp = \mathcal{O}$, i.e. $\wp$ is torsion. But then f is a rational combination of ω_1, ω_2 and $Z = 0$.

$\square$

<u>Corollary III.2.5</u>: $\Lambda f = \Lambda \int_{\mathcal{O}}^{\wp} \frac{dX}{Y} = Z \in K(X)$ is not in the image $\Lambda K(X)$ of the injection $\Lambda: K(X) \to K(X)$ except for $Z = 0$ i.e. $\wp$ torsion.

<u>Proof</u>: If $Z = \Lambda Z'$ for some $Z' \in K(X)$ then f has trivial periods. Note Z' and f both satisfy the inhomogeneous equation and so differ by a complex linear combination of ω_1 and ω_2.

$\square$

We now come to the two crucial results which make all our computations possible. Recall that we have shown that the map

$$E^{gen}(K(X)) \longrightarrow K(X)$$

$$\wp \longrightarrow \Lambda \int_{\mathcal{O}}^{\wp} \frac{dX}{Y} = Z$$

has exactly the torsion as kernel. Thus the image is $\mathbb{Z}^r$ where r is the rank of $E^{gen}(K(X))$, i.e. the rank of the group of sections of E over X. We have the important result:

<u>Theorem III.2.6</u>: (Independence Theorem) The image of this map spans a complex vector subspace of $K(X)$ of rank r.

<u>Proof</u>: See Stiller [29] Theorem 3.8.   □

Thus there is a finite dimensional subspace of dimension r sitting in K(X) viewed as a complex vector space. One would like to find a divisor $\mathcal{A}$ such that this subspace is contained in the linear system $L(\mathcal{A}) = \{f \in K(X) \text{ s.t. } \text{div}(f) + \mathcal{A} > 0\}$. Riemann-Roch then gives an estimate for r.

In order to find $\mathcal{A}$ and make such an estimate we shall make a geometric assumption. Namely, we assume $E^{gen}/K(X)$ has only multiplicative reduction, or equivalently that E/X has only fibers of type I_b (see Kodaira [16]). This assumption on the fibers will hold for most of Parts III and IV. We will consider the general case in V.3. We then have:

<u>Theorem III.2.7</u>: (Existence of Good Local Models) For each point x ε X we can choose a model of E^{gen} over K(X) so that the function f has at least one holomorphic single-valued non-vanishing branch at x. Here $f = \int_{\theta}^{p} \frac{dX}{Y}$ is the multivalued function associated to an element $\mathcal{P}$ of $E_0^{gen}(K(X))$, so as a section it lies in B_0 (see the remarks at the beginning of this section).

<u>Proof</u>: Recall the exact sequence

$$0 \to G \to \mathcal{O}(\mathcal{F}) \to \Omega(B_0) \to 0$$

discussed at the beginning of this section and the related exponential map

together with $\mathcal{P}$ thought of as a section of B_0 over X. The map $\mathcal{F} \xrightarrow{e} B_0$ is surjective, so for any x ε X, $e^{-1}(\mathcal{P})$ <u>must</u> meet the fiber

f_x of f over x. At a good fiber, that is local monodromy $\begin{pmatrix} 1 & 0 \\ 0 & 1 \end{pmatrix}$, the

set $e^{-1}(\varphi)$ is a covering of a neighborhood U of x with "$\mathbb{Z} \oplus \mathbb{Z}$"

branches. In case of local monodromy $\begin{pmatrix} 1 & b \\ 0 & 1 \end{pmatrix}$ b $\epsilon \mathbb{Z}$, b > 0 there are "$\mathbb{Z}$"

points in $e^{-1}(\varphi)$ over x all differing by ng_2, n $\epsilon \mathbb{Z}$, where g_2 is a

generator of the invariant sections of G at x. In all other cases there

would be only one point in $e^{-1}(\varphi)$ over x but we have ruled this out by

assuming multiplicative reduction. Here the section g_2 should not be

confused with the function ω_2 which may degenerate at x.

We shall deal separately with the cases. Firstly, if the local

monodromy at x is $\begin{pmatrix} 1 & 0 \\ 0 & 1 \end{pmatrix}$ then the fiber at x is good, i.e., an

elliptic curve E_x, both with regard to the original surface E/X and with

regard to B_0. Having chosen a model for the generic fiber E^{gen} over

K(X) and a parameter t ϵ K(X) (which we will suppose to be a good local

parameter at x), we arrive at a differential equation as in section 1

above. Now the point x under consideration may be a cosingular point for

this differential equation but as our analysis of the local behavior of

these equations shows (see Part II, §2), we can adjust the model

(multiplying the periods by a suitable rational function) so that the

exponents of the differential equation are (0,s) s $\epsilon \mathbb{Z}$, s $\geq$ 1. Then the

solutions ω_1, ω_2 will be holomorphic non-vanishing single-valued

functions and it will be possible to trivialize f in a neighborhood of x

so as to have the sections of G represented by $m\omega_1 + n\omega_2$ m, n $\epsilon \mathbb{Z}$. For

this trivialization $e^{-1}(\varphi)$ is given by the branches of f on this

neighborhood of x and the result follows. Note f is computed here

using this new model.

In the case where the local monodromy is $\begin{pmatrix} 1 & b \\ 0 & 1 \end{pmatrix}$ b > 0, the fiber is

singular of type I_b b > 0. Again using the analysis of the local

behavior of the differential equation we can adjust the model so that the

exponents are (0,0) and ω_2 will be a holomorphic non-vanishing single-valued function representing the invariant sections of G in a neighborhood of x for a suitable trivialization of $\hat{f}$. For this trivialization $e^{-1}(\varphi)$ is given by the branches of f. The existence of the desired branch then follows from a similar result, Theorem 3.7, in Stiller [29]. □

We remark that a similar result holds without assuming multiplicative reduction but we can no longer say that the single-valued holomorphic branch is non-vanishing and it is more difficult to relate the choices of model for E^{gen} over K(X) to the trivializations of $\hat{f}$.

As a corollary to this result we can show that the functions $Z \in K(X)$ that arise as $\Lambda f = Z$ where $f = \int_{\theta}^{\mathcal{P}} \frac{dX}{Y}$, i.e. the Z in the image of the map

$$E^{gen}(K(X)) \longrightarrow K(X)$$

$$\mathcal{P} \longmapsto \Lambda \int_{\theta}^{\mathcal{P}} \frac{dX}{Y} \; ,$$

satisfy the parabolic residue conditions of Definition II.3.3:

Corollary III.2.8: For $\mathcal{P} \in E^{gen}(K(X))$, we have $\Lambda \int_{\theta}^{\mathcal{P}} \frac{dX}{Y} = Z$ in L_{Λ}^{para}. Moreover $Z \notin L_{\Lambda}^{para} \cap \{\Lambda Z': Z' \in K(X)\}$ unless $Z = 0$, i.e. $\mathcal{P}$ torsion. (See Part II, §3.)

Proof: Let $f = \int_{\theta}^{\mathcal{P}} \frac{dX}{Y}$. Then Proposition III.1.2 gives as usual

$$f = (\int_{x_0}^{x} \frac{-\omega_2 Z}{W} dx + c_1)\omega_1 + (\int_{x_0}^{x} \frac{\omega_1 Z}{W} + c_2)\omega_2 \; .$$

At a cosingular point every branch of f is single-valued as the proof of the theorem shows. Thus the residue condition holds. At any other singular point (true singularity) f will have at least one single-valued branch as the theorem shows. Any other branch is

$$f + \tilde{c}_1 \omega_1 + \tilde{c}_2 \omega_2 \qquad \tilde{c}_1, \tilde{c}_2 \in \mathbb{C}.$$

The periods are thus

$$[m_\gamma, n_\gamma] = [\tilde{c}_1, \tilde{c}_2](M_\gamma - I)$$

where γ is a simple loop around the singular point. Our discussion in Part II, §3 shows this is equivalent to the parabolic residue condition. □

<u>Corollary III.2.9</u>: Every $\wp \in E^{gen}(K(X))$ determines a parabolic cohomology class relative to the monodromy representation of $\pi_1(X_0, x_0) \to SL_2(\mathbb{Z})$ where $X_0 = X - S$, S the set of singular fibers of E/X. Moreover the only $\wp \in E^{gen}(K(X))$ which determine the trivial class are the torsion elements.

<u>Proof</u>: Clear. □

We now return to the problem of finding the divisor. As before we assume $E^{gen}/K(X)$ has only multiplicative reduction.

We first choose a model for E^{gen}:

$$Y^2 = 4X^3 - g_2 X - g_3 \qquad g_2, g_3 \in K(X)$$

and a parameter $x \in K(X)$. We then obtain the K-equation Λ annihilating the periods of $\frac{dX}{Y}$ for this model. The period functions ω_1, ω_2 form a K-basis of Λ as usual. Because of our geometric assumptions Λ has only cosingular points and parabolic points of trace 2 among its singularities. (Compare with Part II, §3.)

In order to avoid trouble caused by the parameter x, we shall compute the parameter invariant quantity $Z(dx)^2$. As we remarked following Lemma III.2.1 above $Z(dx)^2$ will be holomorphic on the set X minus the singular and cosingular points of Λ.

Now consider a cosingular point x of Λ where the local monodromy is $\begin{pmatrix} 1 & 0 \\ 0 & 1 \end{pmatrix}$. At such a point Λ has exponents (r,s), $r < s$, $r, s \in \mathbb{Z}$, and ω_1, $\omega_2 = t^r \cdot$ (holomorphic non-vanishing) in terms of a local parameter t at x. But we know that by changing our model for E^{gen} to one where the exponents at x are $(0, s-r)$, the function f will have a holomorphic non-vanishing single-valued branch. (This by Theorem III.2.7.) It then follows that for this new model $Z(\frac{dx}{dt})^2$ will be holomorphic if $s-r = 1$ and will have at most a first order pole if $s-r > 1$. So for the original model $Z(dx)^2$ has order r if $s-r = 1$ and $r-1$ if $s-r > 1$. The reader may wonder about the difference between $s-r = 1$ and > 1. The explanation is this: because the exponents are distinct integers at the cosingular points there are higher order conditions that must come into play to guarantee no logarithmic behavior in the solutions of the homogeneous equation; thus assuring that the local monodromy will be trivial. For example, in the case of exponents $(0,1)$ the equation must be

$$\frac{d^2}{dt^2} + \left(\frac{0}{t} + \text{holomorphic}\right)\frac{d}{dt} + \left(\frac{0}{t^2} + \frac{q_{-1}}{t} + \text{holomorphic}\right)$$

and the higher order conditions force $q_{-1} = 0$. These conditions in turn impose conditions on Z in the inhomogeneous case so as to assure local single-valued solutions. In the case where $s-r > 1$ we have yet to take into account these higher order conditions, but note that they are clearly equivalent to the residue condition of Definition II.3.2.

For the case of parabolic local monodromy with trace 2 we have that the exponents of Λ are (r,r) $r \in \mathbb{Z}$. We can arrange our K-basis so that

$$\omega_2 = t^r \cdot \text{(holomorphic non-vanishing)}$$

is a basis for the invariant solutions at x. Again using Theorem III.2.7,

we can change the model to one where the exponents are (0,0) and here f

will have a holomorphic non-vanishing single-valued branch. Thus $Z(\frac{dx}{dt})^2$

for this model will have at most a first order pole, and for the original

model $Z(dx)^2$ will have order $r-1$.

We can now summarize these results in the following:

Theorem III.2.10: Given $Y^2 = 4X^3 - g_2X - g_3$ a model for the generic

fiber E^{gen} over $K(X)$ of the elliptic surface E over X, we determine

the K-equation Λ annihilating the periods of $\frac{dX}{Y}$ by the formulas given

previously. Define a divisor $\mathcal{B}$ as follows:

$$
\text{ord}_x\mathcal{B} = \begin{cases} -(r-1) & \text{if } \Lambda \text{ has exponents } (r,s) \text{ with} \\ & \quad s \geq r \text{ at } x \text{ and } s-r \neq 1. \\ \\ -r & \text{if } \Lambda \text{ has exponents } (r,s) \text{ with} \\ & \quad s \geq r \text{ at } x \text{ and } s-r = 1. \end{cases}
$$

Then for any $\wp \in E^{gen}(K(X))$ we have that

$$
\Lambda \int_{\mathcal{O}}^{\wp} \frac{dX}{Y} = Z \in L(\mathcal{B} + \text{div}(dx)^2)
$$

Proof: Some multiple of $\wp$ is in $E_0^{gen}(K(X))$, so the above discussion

applies, and the result then follows.   □

Corollary III.2.11: Let r be the rank of the free part of the

Mordell-Weil group $E^{gen}(K(X))$, then

$$
r \leq \dim L(\mathcal{B} + \text{div}(dx)^2) = \dim L(\mathcal{B} + 2K)
$$

K the canonical divisor on X.   □

Remark: The divisor $\mathcal{B}$ depends on the model of E^{gen}, but if we choose

another model

$$Y^2 = 4X^3 - h^{-4}g_2X - h^{-6}g_3 \qquad h \in K(X)^*$$

then the new equation is the twisted equation Λ_h and

$$\Lambda_h(hf) = hZ.$$

So that then the new divisor is $\mathcal{B} + \text{div}(h)$ which is linearly equivalent to $\mathcal{B}$.

We can obtain a completely invariant description of this map by viewing the image as certain holomorphic sections of $(\Omega_X^1)^{\otimes 2} \otimes \mathcal{L}(\mathcal{B})$, the line bundle of quadratic differentials tensored with $\mathcal{L}(\mathcal{B})$ the line bundle corresponding to the divisor class of $\mathcal{B}$.

We also remark that if E/X has additive reduction then on a suitable ramified cover of X, E has multiplicative reduction. The bound above then applies to give a bound on the original rank r. The case of additive reduction is also discussed in V.3.

The divisor $\mathcal{Cl} = \mathcal{B} + \text{div}(dx)^2$ plays an important role in what follows and the reader should note its definition.

§3. Automorpic forms and a result of Hoyt's.

In the previous sections we discussed a mapping

$$E^{\text{gen}}(K(X)) \longrightarrow K(X)$$

$$\mathcal{P} \longrightarrow \Lambda \int_{\mathcal{O}}^{\mathcal{P}} \frac{dX}{Y} = Z \quad .$$

The function $f = \int_{\mathcal{O}}^{\mathcal{P}} \frac{dX}{Y}$ then satisfies the inhomogeneous differential equation $\Lambda f = Z$ and so it is of the form

$$f = (\int_{x_0}^{x} \frac{-\omega_2 Z}{W} + c_1)\omega_1 + (\int_{x_0}^{x} \frac{\omega_1 Z}{W} + c_2)\omega_2$$

(see Theorem I.2.1 or Proposition III.1.2). Via the integral transforms 4) and 5) of Part I, §3 we can associate to f (or rather $F = f/\omega_2$) an "automorphic form" g of weight 3. Thus to each section of E/X we can associate an "automorphic form" g of weight 3 on the upper-half-plane $\hbar$ viewed as the universal cover of a suitable Zariski open subset $X_0 \subset X$. We will have

$$g(\gamma z) = (c_\gamma \omega(z) + d_\gamma)^3 g(z) \qquad z \in \hbar$$

where $\gamma \in \overline{\Gamma} \subset PSL_2(\mathbb{R})$ corresponds to a path γ in $\pi_1(X_0, x_0)$ via the uniformization, $\omega(z) = \omega_1(z)/\omega_2(z)$, and $\begin{pmatrix} a_\gamma & b_\gamma \\ c_\gamma & d_\gamma \end{pmatrix}$ is the monodromy matrix for ω_1, ω_2 around γ. Moreover we have seen that g is independent of the particular choice of solution to $\Lambda f = Z$, i.e. it is independent of c_1 c_2, as well as the choice of base point $x_0 \in X_0$ provided we choose appropriate branches of ω_1, ω_2 at the new base point. Moreover it is independent of the choice of parameter used to form Λ as well as "twists" of Λ by $h \in K(X)$ to Λ_h (see remarks following Theorem I.3.1). Thus in the case where g comes from a section $\wp$ as above, g is independent of the model chosen for $E^{gen}/K(X)$ and the parameter chosen. It depends only on the choice of a basis for the periods or more precisely on $\omega = \omega_1/\omega_2$ which is determined up the natural action of $SL_2(\mathbb{Z})$.

These automorphic forms were first introduced in Hoyt [12] in the context of elliptic surfaces. In that paper Hoyt proves a number of very important results concerning the automorphic forms attached to the sections of an elliptic surface. In this section we shall reprove these facts by entirely different means, making use of our explicit computations of g, i.e.

$$g(z) = \frac{\omega_2(z)^3 Z(z)}{W(z)^2} \qquad \text{(see Theorem I.3.1)}$$

and of the K-equations Λ which arise.

Let E/X be an elliptic surface which as usual will be assumed to have a section, to have no exceptional curves of the first kind in the fibers, and to have non-constant functional invariant $J \in K(X)$. We will also make the assumption that $E^{gen}/K(X)$ has only multiplicative reduction, i.e. that E/X has only singular fibers of type I_b $b > 0$.

Let $S \subset X$ be the support of the set of singular fibers and $X_0 = X - S$. Consider

$$\pi: \; \hbar, z_0 \to X_0, x_0$$

to be the universal cover with base point z_0 going to x_0, and $\overline{\Gamma} \subset PSL_2(\mathbb{R})$ the group of covering transformations which corresponds to $\pi_1(X_0, x_0)$.

We shall begin by determining the so called q-expansion of g at the cusps of $\overline{\Gamma}$. These cusps correspond to the points of $S \subset X$ where E has singular fibers of type I_b $b > 0$. Let $s \in S$ be such a point. After conjugation if necessary we can assume that $i\infty$ is the cusp we are concerned with lying over $s \in S \subset X$.

Let $\Lambda_{(J, \lambda)}$ be the K-equation associated to a particular model of $E^{gen}/K(X)$ and a particular choice of parameter $x \in K(X)$. We fix a K-basis ω_1, ω_2 corresponding to the periods of $\frac{dX}{Y}$ for this model so that the period map ω sends $i\infty$ to $i\infty$

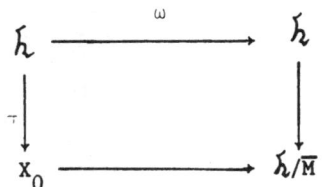

$\overline{M}$ the global projective monodromy group in $PSL_2(\mathbb{Z})$.

Thus $\begin{pmatrix} \omega_1 \\ \omega_2 \end{pmatrix} \to \begin{pmatrix} 1 & b \\ 0 & 1 \end{pmatrix} \begin{pmatrix} \omega_1 \\ \omega_2 \end{pmatrix}$ around $s \in S$ where the singular fiber is of type I_b, i.e. J has a pole of order b, $b \geq 1$. Setting $q = e^{2\pi i z/h}$ gives us a good local parameter at s where $\gamma_h = \pm\begin{pmatrix} 1 & h \\ 0 & 1 \end{pmatrix}$ $h > 0$ mod $\pm\begin{pmatrix} 1 & 0 \\ 0 & 1 \end{pmatrix}$ is the positive translation of $\hslash$ which generates the isotropy group of $i\infty$ in $\bar{\Gamma}$. Note that $g(\gamma_h z) = g(z)$ so we have a Fourier series expansion.

__Theorem III.3.1:__ $g(z) = \sum_{n>0} a_n e^{2\pi i n z/h} = \sum_{n>0} a_n q^n$. Thus g is holomorphic and vanishes at the cusps.

__Proof:__ Because ω_2 is invariant around s it suffices to simply compute locally on X at s and determine the order of $\dfrac{\omega_2(x)^3 Z(x)}{W(x)^2}$ which is single-valued at s. Moreover the exponents of $\Lambda_{(J,\lambda)}$ at s must be (r,r) $r \in \mathbb{Z}$ and since g is independent of the model we can assume without loss of generality that the exponents are $(0,0)$. As g is independent of the parameter we can assume x is a good local parameter at s (take it to be q if you like). Then:

1) ω_2 is holomorphic non-vanishing (see Part II, §2)

2) Z has at most a first order pole (see Theorem III.2.8)

3) $W = \lambda^2 \dfrac{\frac{dJ}{dx}}{J}$ up to a constant multiple. But $\lambda^2 \in K(X)$ must be non-vanishing at s because the exponents are $(0,0)$ (see Part II, §2) and $\dfrac{\frac{dJ}{dx}}{J}$ has a first order pole.

Thus $g = \dfrac{\omega_2^3 Z}{W^2}$ is holomorphic and vanishing. $\qquad\qquad\qquad$ □

To consider the behavior of $g(z)$ on $\hslash$ itself we need to reflect for a moment on the possible poles of g. Over X_0 our surface has no singular fibers. Thus $\Lambda_{(J,\lambda)}$ has at most cosingular points on X_0 and at such points the exponents of $\Lambda_{(J,\lambda)}$ will be (r,s) r, $s \in \mathbb{Z}$, $s-r \geq 1$. But because g is independent of the model and the choice of parameter, we can always arrange by a change of model that the exponents be $(0, s-r)$

and that we have a good local parameter at such a point. It follows that
if s−r = 1 then the equation is holomorphic and in terms of the good
local parameter ω_2, W are holomorphic invertible functions and Z is
holomorphic so that g is holomorphic.

We can see from our local computations in Part II, §2 that the
meromorphic behaviour occurs at points where ord dJ $\geq$ 1. Notice that at
points where J = 0 the zero must be of order ≡ 0 mod 3 and where J = 1
the one must be of order ≡ 0 mod 2 otherwise we would necessarily have a
point with additive reduction. If J ≠ 0, 1 and ord dJ = n $\geq$ 1 then the
exponent difference is n+1 as we have seen from our local calculations in
Part II, §2 (λ^2 plays no role in the exponent difference but only adjusts
both exponents up or down by the same element of (1/2) $\mathbb{Z}$). If J = 0 to
order 3n n $\geq$ 1 then the exponent difference is n. (So for example if
J = 0 to order 3 the exponent difference is 1 and after adjusting the
model (which is the same as twisting the differential equation) we can
assume the exponents are (0,1) so that λ^2 = t · (holomorphic
non−vanishing). Then in terms of the good local parameter the Wronskian is

$$W = \lambda^2 \frac{\frac{dJ}{dt}}{J}$$

which is holomorphic non−vanishing as is ω_2. Since we have seen Z must
be holomorphic in this case (see §2 above), we have g holomorphic.) When
J = 1 to order 2n n $\geq$ 1 then the exponent difference is n and in the
special case n = 1 where the exponents can be taken as (0,1) after a
change of model we get g holomorphic. Let

$$k = \begin{cases} n+1 & \text{if } J \neq 0,1 \text{ but } \text{ord } dJ = n \geq 1. \\ n & \text{if } J = 0 \text{ to order } 3n \text{ } n \geq 1. \\ n & \text{if } J = 1 \text{ to order } 2n \text{ } n \geq 1. \\ 1 & \text{otherwise .} \end{cases}$$

Thus k is exactly the order of ramification (or the order of ramification divided by 3 or 2) of the map $X \xrightarrow{J} P^1_{\mathbb{C}}$, and it is equal to the exponent difference.

Theorem III.3.2: Let $z \in \mathcal{h}$ be a point lying over a point $x \in X_0$ then

$$g(z) = \sum_{n \geq -2k+1} a_n z^n$$

and $a_{-1} = 0$ if $k = 1$ i.e. if the exponent difference is 1. Thus $g(z)$ is holomorphic when the exponent difference is 1.

Proof: Without loss of generality we can compute in neighborhood of $\pi(z) = x \in X_0$ and can assume the model is chosen so that the exponent difference is $(0,k)$ k being as above and that t, a good local parameter at x, is being used to form Λ. We then have:

1) ω_2 is holomorphic non-vanishing

2) Z has a pole of order at most 1 if $k > 1$ and is holomorphic otherwise

3) W has a zero of order exactly $k-1$.

and the result follows.   □

Remark: Theorems III.3.1 and III.3.2 give us Hoyt's original result except for one additional point. Hoyt shows that for a suitable parameter one can arrange that in the expansion for g one will have $a_{-k} = 0$. For $k = 1$ we have already seen that this is true. And we would like to discuss the source and nature of this extra condition.

    At any point $x \in X_0$ not a true singular point, in particular at a cosingular point, the fiber at x is good, in other words an elliptic

curve. It follows then that for a given $\mathcal{P} \in E^{gen}(K(X))$ the resulting

function

$$\oint_{\theta}^{\mathcal{P}} \frac{dX}{Y} = f = (\int_{x_0}^{x} \frac{-\omega_2 Z}{W} + c_1)\omega_1 + (\int_{x_0}^{x} \frac{\omega_1 Z}{W} + c_2)\omega_2$$

used in determining the "automorphic form" g has zero periods around x.

This fact is of course independent of c_1, c_2 because the local monodromy

matrix is the identity. Thus the condition comes down to

$$\frac{-\omega_2 Z}{W} dx \qquad \text{and} \qquad \frac{\omega_1 Z}{W} dx$$

having no residue, which is the residue condition of Part II, §3, and in

fact the parabolic residue conditions hold because

$$\frac{\tilde{\omega} Z}{W} dx$$

is holomophic at the cusps where $\tilde{\omega}$ is an invariant solution (this can be

seen from the proof of Theorem III.3.1 above).

In general then when the exponents are (0,k) $k \geq 2$ an additional

linear condition is imposed on Z to insure that we have no residue. One

can easily verify that this is equivalent to Hoyt's condition.

Let

$T_3(\mathcal{O}\mathcal{U}) = \{$meromorphic functions g on $\mathcal{H}$ satisfying the

transformation law $g(\gamma z) = (c_\gamma \omega + d_\gamma)^3 g(z)$ as usual, subject

to the local conditions given in Theorems III.3.1 and

III.3.2.$\}$

and let

$T_3^{para}(\mathcal{O}) \subset T_3(\mathcal{O})$ be the set satisfying the additional linear

conditions, as described above or as in Hoyt [12], which are

equivalent to the parabolic residue conditions. Thus

$T_3^{para}(\mathcal{O})$ consists of the automorphic forms of the second kind

in $T_3(\mathcal{O})$ (see Part III, §5 below).

Here $\mathcal{O}$ is the divisor defined by $\mathcal{B} + \mathrm{div}(dx)^2$ as in Theorem III.2.10.
We define

$$L(\mathcal{O}) = \{Z \in K(X): \mathrm{div}(Z) + \mathcal{O} \geq 0\}$$

and

$L_\Lambda^{para}(\mathcal{O}) \subset L(\mathcal{O})$ the sub-linear system consisting of those Z

satisfying the parabolic residue conditions.

We then have:

__Theorem III.3.3:__ The linear map

$$T_3(\mathcal{O}) \xleftarrow{\quad L_3 \quad} L(\mathcal{O})$$

$$g = \frac{\omega_2^3 Z}{w^2} \longleftarrow Z$$

is an isomorphism under which $T_3^{para}(\mathcal{O})$ is isomorphic to $L_\Lambda^{para}(\mathcal{O})$. $\square$
This gives a comparison between the results of Hoyt and our results above.
Namely the following diagram commutes:

$$E^{gen}(K(X)) \longrightarrow T_3^{para}(\mathcal{O}) \subset T_3(\mathcal{O}) \subset \text{"forms of weight 3"}$$

$$L_\Lambda^{para}(\mathcal{O}) \subset L(\mathcal{O}) \subset K(X)$$

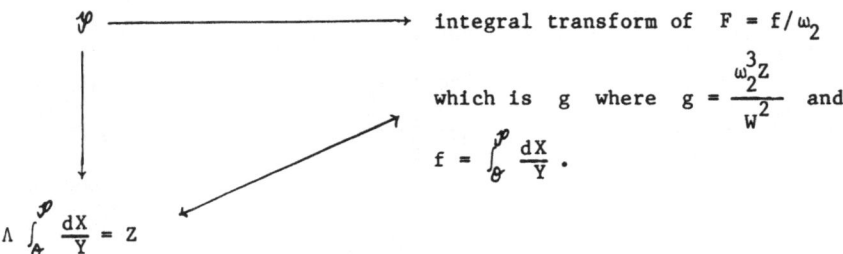

$\mathcal{P}$ —————————→ integral transform of $F = f/\omega_2$

which is g where $g = \dfrac{\omega_2^3 Z}{W^2}$ and

$f = \displaystyle\int_{\mathscr{S}}^{\mathcal{P}} \dfrac{dX}{Y}$.

$\Lambda \displaystyle\int_{\mathscr{S}}^{\mathcal{P}} \dfrac{dX}{Y} = Z$

We now have the important result first due to Hoyt in the case of the automorphic forms attached to the sections of E/X.

<u>Theorem III.3.4</u>: An element of $T_3^{para}(\mathcal{O})$ corresponds to a section if and only if it has integer periods. Equivalently $Z \in L_\Lambda^{para}(\mathcal{O})$ comes from a section if and only if for some choice of c_1, c_2

$$f = \left(\int_{x_0}^{x} \frac{-\omega_2 Z}{W} + c_1\right)\omega_1 + \left(\int_{x_0}^{x} \frac{\omega_1 Z}{W} + c_2\right)\omega_2$$

has integer periods. Thus for every $\gamma \in \pi_1(X_0, x_0)$

$$f \;\to\; f + m_\gamma \omega_1 + n_\gamma \omega_2 \qquad m_\gamma, \, n_\gamma \in \mathbb{Z}$$

where $X_0 = X - S$, S being the support of the singular fibers. This makes sense because $Z \in L_\Lambda^{para}(\mathcal{O})$ assures us of zero periods around any other point.

<u>Proof</u>: Clearly for $Z \in L_\Lambda^{para}(\mathcal{O})$ with integer periods we get a section defined over a suitable Zariski open subset. It is enough to see that f has a holomorphic single-valued branch when we choose a good local parameter and adjust the model so that the exponents are $(0,s)$ $s \geq 0$. Such an f will give a "holomorphic section" of $\mathcal{f}$ which exponentiates to give the desired section. This is relatively straightforward and we omit it here.   □

§4. Periods and the rank of $E^{gen}(K(X))$.

As usual we consider an elliptic surface E over a curve X. We assume that E has a section, has no exceptional curves in the fibers, and has multiplicative reduction, i.e. singular fibers of type I_b only. Let S be the support of the singular fibers which is non-empty since the functional invariant $J \in K(X)$ is assumed non-constant. Let $X_0 = X - S$. Choose a model

1) $Y^2 = 4X^3 - g_2 X - g_3$ $g_2, g_3 \in K(X)$

for the generic fiber E^{gen} over $K(X)$. The Mordell-Weil group $E^{gen}(K(X))$ of the $K(X)$-rational points on E^{gen} is a finitely generated abelian group whose rank we denote by r. Recall the injective map

2) $$E^{gen}(K(X))/\text{torsion} \longrightarrow K(X)$$
$$\wp \ \text{mod torsion} \longrightarrow \Lambda \int_{\mathscr{O}}^{\mathscr{P}} \frac{dX}{Y} = Z$$

where $\Lambda = \Lambda_{(J,\lambda)}$ is the K-equation for the model 1) taken with respect to some parameter $x \in K(X)$.

In section 2 above we found a divisor $\mathcal{O}$ so that the image of the map 2) lies in $L(\mathcal{O})$ and in section 3 we saw that in fact this image lies in a sub-linear system $L_\Lambda^{para}(\mathcal{O})$ of the complete linear system $L(\mathcal{O})$ where $L_\Lambda^{para}(\mathcal{O})$ is determined by the parabolic residue conditions.

–Theorem III.4.1: The rank r of the free part of $E^{gen}(K(X))$, the Mordell-Weil group, is less than the dimension of

$$L_\Lambda^{para}(\mathcal{O}) \ / \ \Lambda K(X) \cap L_\Lambda^{para}(\mathcal{O}).$$

Proof: Let X have genus g and assume the cardinality of S is #S ≥ 1. Then $\pi_1(X_0, x_0)$ is a free group on $2g + \#S - 1$ generators. Now given $Z \in L_\Lambda^{para}(\mathcal{OL})$ we can form the function

$$f = \left(\int_{x_0}^{x} \frac{-\omega_2 Z}{W} \, dx + c_1\right)\omega_1 + \left(\int_{x_0}^{x} \frac{\omega_1 Z}{W} \, dx + c_2\right)\omega_2$$

which is a solution of the inhomogeneous equation

$$\Lambda f = Z.$$

The periods of f around $\gamma \in \pi_1(X_0, x_0)$ are given by the analytic continuation around γ:

$$f \rightarrow f + m_\gamma \omega_1 + n_\gamma \omega_2 \qquad m_\gamma, n_\gamma \in \mathbb{C}.$$

The residue condition on Z guarantees that we do not have periods around cosingular points but only around true singularities.

Proposition I.4.5 shows that the periods are completely determined by their value around the generators of $\pi_1(X_0, x_0)$. Thus we have a period map Φ which is a $\mathbb{C}$-linear map:

$$L_\Lambda^{para}(\mathcal{OL}) \xrightarrow{\Phi} \mathbb{C}^{4g-2+2(\#S)}/V$$

where V is the image of

$$\mathbb{C}^2 \longrightarrow \mathbb{C}^{4g-2+2(\#S)}$$

$$[c_1, c_2] \longrightarrow \bigoplus_\gamma [c_1, c_2](M_\gamma - I)$$

and gives the natural equivalence on periods caused by adjusting c_1, c_2. (This is adjusting by a parabolic coboundary in parabolic cohomology as mentioned in Part II, §3.)

Clearly V has dimension 2 as the global monodromy has finite index in $SL_2(\mathbb{Z})$.

The kernel of Φ is $\Lambda K(X) \cap L_{\Lambda}^{para}(\mathcal{O}l)$ by Theorem I.4.3. On the other hand by Corollary III.2.4 if Z is of the form

$$Z = \Lambda \int_{\mathcal{O}}^{\mathcal{P}} \frac{dX}{Y}$$

for some $\mathcal{P} \in E^{gen}(K(X))$ then the function

$$f = \int_{\mathcal{O}}^{\mathcal{P}} \frac{dX}{Y}$$

never has trivial periods unless $\mathcal{P}$ is torsion. Thus we have maps

$$E^{gen}(K(X))/\text{torsion} \longrightarrow L_{\Lambda}^{para}(\mathcal{O}l)/\Lambda K(X) \cap L_{\Lambda}^{para}(\mathcal{O}l) \xrightarrow{\Phi} \mathbb{C}^{4g-2+2(\#S)}/V$$

which are injective and the result follows. $\square$

Theorem III.3.3 says that an element of $L_{\Lambda}^{para}(\mathcal{O}l)$ with integer periods comes from a section so to determine r we must make a more detailed analysis of Φ .

Let's consider then a small loop γ around a point $s \in S$ with γ one of our generators. Now the periods are completely independent of the model and the parameter, so we assume Λ has exponents $(0,0)$ at s and that t is a good parameter. Then

$$f = \left(\int_{x_0}^{x} \frac{-\omega_2 Z}{W} \, dt + c_2\right)\omega_1 + \left(\int_{x_0}^{x} \frac{\omega_1 Z}{W} \, dt + c_2\right)\omega_2$$

Choose branches of ω_1 , ω_2 at the base point x_0 so that ω_2 is invariant around s. We then have locally at s that

$$\omega_2 = a_0 + a_1 t + \dots \qquad\qquad a_0 \neq 0$$

$$\omega_1 = \omega_2 \left(\frac{b}{2\pi i} \log t + \text{holomorphic} \right)$$

where the singular fiber is of type I_b $b > 0$. It follows that

$$\frac{-\omega_2 Z}{W} \, dt \quad \text{is holomorphic}$$

and

$$\frac{\omega_1 Z}{W} \, dt \quad \text{has only logarithmic growth.}$$

Thus using the results in Part I concerning the periods we have

$$[m_\gamma, n_\gamma] = [0, n_\gamma] = [0, \chi_2(\gamma)] + [0, bc_1]$$

(see Proposition I.4.2). We also have the conditions of Proposition I.4.4
holding with $N < 1$. So by Proposition I.4.9

$$[m_\gamma, n_\gamma] = [0, b(\chi_1(\gamma_1) + c_1)]$$

where γ_1 is a direct path from x_0 to s corresponding to γ. Thus

$$m_\gamma = 0$$

$$n_\gamma = b \left(\int_{x_0}^{s} \frac{-\omega_2 Z}{W} \, dt + c_1 \right)$$

for an appropriate path from x_0 to s. This means that the image of Φ

$$L_\Lambda^{\text{para}}(\theta) / \Lambda K(X) \cap L_\Lambda^{\text{para}}(\theta) \xrightarrow{\Phi} \mathbb{C}^{4g-2+2(\#S)}/V$$

has dimension at most $4g - 4 + (\#S)$. Note that if E has geometric genus
$p_g = 0$, it then follows that

$$r = \dim L_\Lambda^{\text{para}}(\theta) / \Lambda K(X) \cap L_\Lambda^{\text{para}}(\theta)$$

since it is well known in that case that

$$r = 4g - 4 + (\#S).$$

Another interesting case is when E is the canonical elliptic surface over a modular curve

$$X = \mathring{h}^{*}/\overline{\Gamma} \qquad \overline{\Gamma} \subset PSL_2(\mathbb{Z})$$

coming from $\Gamma \subset SL_2(\mathbb{Z})$ having finite index, no elements of finite order, and only regular cusps (see Shioda [27]). In this case:

$$T_3^{para}(\mathcal{O}\!\ell) = T_3(\mathcal{O}\!\ell) = \text{ordinary space of cusp forms} \quad S_3(\Gamma)$$
$$\text{of weight 3 with respect to} \quad \Gamma.$$

Shioda has shown that for these E the rank r of the Mordell–Weil group is 0 and that $\dim_{\mathbb{C}} S_3(\Gamma) = p_g$. Suppose $g \in T_3^{para}(\mathcal{O}\!\ell)$ has zero periods, i.e. the corresponding $Z \in L_\Lambda^{para}(\mathcal{O}\!\ell)$ lies in $\Lambda K(X) \cap L_\Lambda^{para}(\mathcal{O}\!\ell)$ then

$$F = f/\omega_2 = \left(\int_{x_0}^x \frac{-\omega_2 Z}{W} \, dx + c_1 \right)\omega + \left(\int_{x_0}^x \frac{\omega_1 Z}{W} \, dx + c_2 \right)$$

is holomorphic on $\mathring{h}$ and at the cusps. We see that F transforms as:

$$F(\gamma z) = (c_\gamma z + d_\gamma)^{-1} F(z) \qquad \gamma \in \Gamma,$$

so that F is an automorphic form of weight -1, of which none exist. Thus

$$0 \leq \dim L_\Lambda^{para}(\mathcal{O}\!\ell) \, / \, \Lambda K(X) \cap L_\Lambda^{para}(\mathcal{O}\!\ell) = \dim S_3(\Gamma) = p_g$$

and $p_g = 2g - 2 + t_1/2$, where g is the genus of X and t_1 the number of cusps of Γ.

§5. A generalization of a result of Shioda's.

Let $\Gamma \subset SL_2(\mathbb{Z})$ be any subgroup of finite index which does not contain $\begin{pmatrix} -1 & 0 \\ 0 & -1 \end{pmatrix}$. Using a construction of Shioda [27] one can construct a canonical basic elliptic surface $E = E_\Gamma$ over the modular curve $X = X_\Gamma \cong \hslash^*/\Gamma$. To any such basic elliptic surface (see Part III, §1) we can associate a K-equation Λ which is unique up to twist and in the case of an elliptic modular surface we can select a K-basis ω_1, ω_2 unique up to a scalar by choosing it so that the monodromy representation is the identity map on Γ. Having done this for $\Gamma \subset SL_2(\mathbb{Z})$ as above we make a definition:

Definition III.5.1: Let g be a meromorphic automorphic form of weight 3 with respect to $\Gamma \subset SL_2(\mathbb{Z})$ in the classical sense. We say g is of the second kind if $g = \omega_2^3 Z/W^2$ for some $Z \quad L_\Lambda^{para}$ and exact if $Z = \Lambda Z'$ for some $Z' \in K(X)$. □

As these notions do not depend on twisting Λ, we see that they depend only on Γ itself, and we have well-defined intrinsic notions of second kind and exact.

We have seen that Z (or f or g) gives the trivial parabolic cohomology class if and only if $Z = \Lambda Z'$ for some $Z' \in K(X)$ (i.e. g is exact). We in fact have isomorphisms

$$\frac{L_\Lambda^{para}}{\Lambda K(X)} \cong \frac{\text{automorphic forms of second kind}}{\text{exact}} \cong H_p^1(\Gamma, V_{\mathbb{C}}^1) \quad (\cong H^{para})$$

where H_p^1 is the complex parabolic cohomology and $V_{\mathbb{C}}^1$ denotes the standard or identity representation of Γ. Among the forms of the second kind are those of the "first kind", i.e. the usual cusp forms of weight 3,

$S_3(\Gamma)$, none of which are exact. We can interpret the isomorphism above as a case of the well-known Shimura isomorphism (see Shimura [26], Bayer and Neukirch [2], or Stiller [36]).

Shioda has shown:

Theorem III.5.2: (Shioda [27]) The space $S_3(\Gamma)$ of cusp forms of weight three with respect to Γ is canonically isomorphic (over $\mathbb{C}$) to the space of holomorphic 2-forms on E_Γ, i.e.

$$S_3(\Gamma) \cong H^0(E_\Gamma, \Omega^2_{E_\Gamma}).$$ □

This discussion can be generalized to arbitrary K-equations Λ so as to give notions of second kind and exact for the generalized automorphic forms constructed above. However caution is called for, especially in selecting the correct notion of first kind (see below).

For example, assume Λ is any K-equation on a curve X whose only true singularities are parabolic with trace +2. (This is equivalent to the corresponding elliptic surface having only singular fibers of type I_b $b > 0$ i.e. multiplicative reduction.) We then have:

Definition III.5.3: Let Λ, ω_1, ω_2 be the K-equation above with fixed K-basis on the curve X. Let S, X_0, $x_0 \in X_0$, $z_0 \in \hbar$, $\overline{\Gamma} \subset PSL_2(\mathbb{R})$ giving $(\hbar/\overline{\Gamma}, z_0) \cong (X_0, x_0)$, and M the monodromy be as above. A meromorphic function $g(z)$ on the upper half-plane $\hbar$ which transforms as

$$g(\gamma z) = (c_\gamma \omega(z) + d_\gamma)^3 g(z)$$

for $\gamma \in \overline{\Gamma}$ corresponding to $\gamma \in \pi_1(X_0, x_0)$, $\begin{pmatrix} a_\gamma & b_\gamma \\ c_\gamma & d_\gamma \end{pmatrix} \in SL_2(\mathbb{Z})$ the monodromy around γ, and which is meromorphic at the cusps (this makes sense because of the local monodromy assumption) will be called a generalized automorphic form of weight 3 and second kind with respect to Λ, ω_1, ω_2, if

$$g(z) = \frac{\omega_2(z)^3 Z(z)}{W(z)^2}$$

for $Z \in L_\Lambda^{para}$ and exact if $Z \in \Lambda K(X)$.

Again this definition is not dependent on twists of Λ, and in fact it only depends on the elliptic surface over X which is associated to Λ, ω_1, ω_2 (see Stiller [28]). The resulting "Shimura isomorphism" and its relationship with the Hodge theory of E over X is discussed in Part IV.

In the remainder of this section we shall show how to extend Shioda's result (Theorem III.5.2) by making use of the generalized automorphic forms discussed above. For simplicity we will consider a basic elliptic surface E having only multiplicative reduction, i.e. only singular fibers of type I_b $b \geq 1$, over a curve X. As usual we let $S \subset X$ denote the finite set of points where E over X has singular fibers. The set S will also be the set of true singularities of any K-equation on X representing the Gauss–Manin connection or Picard–Fuchs equation of E over X. Let Λ be such a K-equation and ω_1, ω_2 a K-basis associated to E/X (so $J(\omega_1/\omega_2) = $ the functional invariant of E (Kodaira [16]) where J is the elliptic modular function). We then have the usual diagram:

$$
\begin{array}{ccc}
z_0 \in \mathcal{h} & \xrightarrow{\;\omega\;} & \mathcal{h} \\[2mm]
\pi \downarrow & & \downarrow \\[2mm]
x_0 \in X_0 = X\text{-}S & \xrightarrow{[\omega]} & C_0 = \mathcal{h}/\overline{M}
\end{array}
$$

where x_0 is a base point in $X_0 = X\text{-}S$, π is the universal covering map with $\pi(z_0) = x_0$, and $\overline{M} \subset PSL_2(\mathbb{Z})$ is the projective monodromy group given by analytic continuation of our chosen branches of ω_1, ω_2 at x_0 around paths $\gamma \in \pi_1(X_0, x_0)$. (See Part I, §1 and Part II, §3 for additional details.) We identify $\pi_1(X_0, x_0)$ with a subgroup $\overline{\Gamma} \subset PSL_2(\mathbb{R})$ in the usual manner. We then have

$$\omega(\gamma z) = M_\gamma z$$

where $\gamma \in \overline{\Gamma}$ corresponds to a path $\gamma \in \pi_1(X_0, x_0)$ and M_γ is the monodromy matrix around that path. An important observation is that because we have assumed that all the singular fibers are of type I_b $b > 0$, if we take a parabolic element $\gamma \in \overline{\Gamma}$ the resulting monodromy matrix M_γ is also parabolic. When we find it necessary to analyze the behavior at some point of S (i.e. cusp of $\overline{\Gamma}$) we will frequently conjugate $\overline{\Gamma}$ and $\overline{M}$ (by choosing another associated K-basis ω_1, ω_2) so that ω carries the cusp at "$i\infty$" to the cusp at "$i\infty$". In other words, we can arrange that going around $s \in S$ once in a counterclockwise manner corresponds to $\pm \begin{pmatrix} 1 & h \\ 0 & 1 \end{pmatrix}$ $h > 0$ mod $\pm \begin{pmatrix} 1 & 0 \\ 0 & 1 \end{pmatrix}$ in Γ, i.e. translation by $h > 0$, and that the monodromy matrix is $\begin{pmatrix} 1 & b \\ 0 & 1 \end{pmatrix}$ if the fiber type is I_b at s.

In order to generalize Shioda's result for elliptic modular surfaces we need an analogue for the space $S_3(\Gamma)$ of cusp forms of weight 3 with respect to Γ.

Recall that a generalized automorphic form of weight three with respect to Λ, ω_1, ω_2 for $\overline{\Gamma} \subset PSL_2(\mathbb{R})$ is a meromorphic function $g(z)$ on $\mathcal{h}$ which satisfies the transformation law

$$g(\gamma z) = (c_\gamma \omega(z) + d_\gamma)^3 g(z) \qquad \omega(z) = \omega_1(z)/\omega_2(z)$$

for $\gamma \in \overline{\Gamma} \subset PSL_2(\mathbb{R})$ and $\begin{pmatrix} a_\gamma & b_\gamma \\ c_\gamma & d_\gamma \end{pmatrix} = M_\gamma \in SL_2(\mathbb{Z})$ the corresponding monodromy matrix. In addition at any cusp of $\overline{\Gamma}$, $g(z)$ must be meromorphic. This last condition makes sense because of our local monodromy assumption; namely without loss of generality assume the cusp is $i\infty$ and that ω carries $i\infty$ to $i\infty$, then if $z \to z + h$ is the minimum positive translation generating the isotropy group of $i\infty$ in $\overline{\Gamma}$ we have

$$g(z + h) = g(z)$$

because $M_\gamma = \begin{pmatrix} 1 & b \\ 0 & 1 \end{pmatrix}$ $b > 0$. We then have a Laurent development for g:

$$g(z) = \sum_{-\infty}^{\infty} a_n e^{2\pi i n z/h} = \sum_{-\infty}^{\infty} a_n q^n \qquad q = e^{2\pi i z/h} .$$

We say that g is meromorphic at the cusp if this expansion has only a finite number of negative terms. (Note this definition is easily extended to handle fiber types I_b^* $b > 0$.)

Definition III.5.4: We shall say that a generalized automorphic form of weight three with respect to Λ, ω_1, ω_2 for $\overline{\Gamma}$ is of the <u>first kind</u> if:

1) $g(z)$ is holomorphic on $\hbar$ except perhaps at the points in $\hbar$ which lie over the cosingular points of Λ in $X_0 = X-S$.

2) At any point $z \varepsilon \hbar$ which lies over a cosingular point where the exponent difference is n, $g(z)$ has a pole of order at most $n-1$. (Actually we can put 1) and 2) together since at a holomorphic point of Λ the exponent difference is 1. In this form we see that the condition is independent of twisting Λ since this only shifts exponents, preserving the difference.)

3) At any cusp lying over $s \varepsilon S$ the Fourier expansion of $g(z)$ has the form:

$$\sum_{n=1}^{\infty} a_n e^{2\pi i n z/h}$$

so that g is holomorphic and vanishes at the cusp.

We denote the space of forms of the first kind and weight 3 by $S_3^1(\overline{\Gamma})$. □

Note that in the elliptic modular case Λ has no cosingular points where

the exponent difference is ≥ 2 so that $S_3^1(\overline{\Gamma})$ is the ordinary space of

cusp forms $S_3(\Gamma)$ as above (assuming $\Gamma \subset SL_2(\mathbb{Z})$ has no torsion so that

$h/\Gamma \cong X$ with $\pi_1(X_0, x_0)$ corresponding to Γ).

Theorem III.5.5: The space $S_3^1(\overline{\Gamma})$ of generalized automorphic forms of the

first kind and weight 3 with respect to Λ, ω_1, ω_2 for $\overline{\Gamma}$ is a subspace

of those of the second kind (see Definition III.5.3).

Proof: Recall that $g(z)$ will be of the second kind if we can write

$$g(z) = \frac{\omega_2(z)^3 Z(z)}{W(z)^2}$$

with $Z \in L_\Lambda^{para}$ (see above). Let $z \in h$ lie over $x \in X_0$ and suppose

that the exponents of Λ at x are $r, s \in \mathbb{Z}$ with $s > r$ and

$s-r = n \geq 1$. In terms of a local parameter t at x we have:

$$g(t) = \frac{a_{-n+1}}{t^{n-1}} + \text{higher order}$$

$$\omega_2(t) = t^r \cdot (\text{holomorphic non-vanishing})$$

$$W(t) = t^{s+r-1} \cdot (\text{holomorphic non-vanishing}).$$

If we consider $\dfrac{g(z) W(z)^2}{\omega_2(z)^3}$ we have a $\overline{\Gamma}$-invariant function on h

which, because g is assumed to be of the first kind and because of the

growth behavior of ω_2, is a rational function Z on X, i.e. $Z \in K(X)$.

Moreover at $x \in X_0$ as above we have that

$$Z(t) = t^{s-1} \text{ (holomorphic)}$$

and

$$\frac{\omega_i(t)Z(t)}{W(t)} dt \text{ is holomorphic for } i = 1,2$$

which means Z satisfies the residue condition of Definition II.3.2.

Only the residue conditions at the cusps remain to be checked. Let $s \in S$ be the point on X to which our cusp corresponds. Without loss of generality we can assume that the cusp is the one at $i\infty$ and that ω carries $i\infty$ to $i\infty$ so that then ω_2 is a local invariant solution at s. Here Λ will have exponents $(r,r) \in \mathbb{Z}$. In terms of a local parameter t ($q = e^{2\pi i z/h}$ for example) at s we have

$$g(t) = t \cdot \text{(holomorphic)}$$

$$\omega_2(t) = t^r \cdot \text{(holomorphic non-vanishing)}$$

$$W(t) = t^{2r-1} \cdot \text{(holomorphic non-vanishing)}.$$

It follows that

$$Z(t) = t^{t-1} \cdot \text{(holomorphic)}$$

and

$$\frac{\omega_2(t)Z(t)}{W(t)} dt \text{ is holomorphic.}$$

Thus $Z \in L_\Lambda^{para}$ as desired. $\square$

The proof actually shows more. Let $\mathcal{O}\!\ell$ be the divisor $\mathcal{B} + \mathrm{div}(dx)^2$ as in Theorem III.2.10 and $L_\Lambda^{para}(\mathcal{O}\!\ell) \subset L(\mathcal{O}\!\ell) = \{Z \in K(X): \mathrm{div}(Z) + \mathcal{O}\!\ell \geq 0\}$ be those Z which are also in L_Λ^{para}, i.e. $L_\Lambda^{para}(\mathcal{O}\!\ell) = L(\mathcal{O}\!\ell) \cap L_\Lambda^{para}$. We then have

Corollary III.5.6: $\quad S_3^1(\overline{\Gamma}) \subseteq T_3^{para}(\mathcal{O}\!\ell) \stackrel{\sim}{=} L_\Lambda^{para}(\mathcal{O}\!\ell)$

$$g = \frac{\omega_2^3 Z}{W^2} \leftrightarrow Z.$$

Proof: Recall that $T_3^{para}(\mathcal{A}) = \{$generalized automorphic forms of weight 3 and of the second kind, subject to the conditions given in Theorem III.3.1 and III.3.2$\}$. The result is then clear. $\qquad\qquad\qquad\qquad\qquad\qquad\qquad$ $\square$

Theorem III.5.7: Let $g \in S_3^1(\overline{\Gamma})$ then g is not exact.

Proof: Recall that g is exact if when we express it as

$$g = \frac{\omega_2^3 Z}{W^2}$$

the function Z is equal to $\Lambda Z'$ for some $Z' \in K(X)$. We need to determine the order of zero or pole of a potential Z' at each point in X. If $x \in X$ is any point, let t be a local parameter at x. In terms of t we have

$$\Lambda Z'(t) = Z(t)\left(\frac{dx}{dt}\right)^2$$

where Λ is $\frac{d^2}{dt^2} + \left(P\,\frac{dx}{dt} - \frac{d}{dt}\log\frac{dx}{dt}\right)\frac{d}{dt} + Q\left(\frac{dx}{dt}\right)^2$. (Recall that Λ is originally given as $\frac{d}{dx^2} + P\frac{d}{dx} + Q$ with $x \in K(X)$ giving a local parameter almost everywhere.) If the exponents of Λ at x are (r,s) with $r, s \in \mathbb{Z}$ and $s \geq r$, then by adjusting the model of E^{gen} over $K(X)$, or equivalently twisting Λ, we can assume $r = 0$, i.e.

$$\Lambda_{t^{-r}}\, t^{-r}Z'(t) = t^{-r}Z(t)\left(\frac{dx}{dt}\right)^2 \,.$$

Because $Z \in L(\mathcal{A})$ by Corollary III.5.6 above we can use the calculations of Theorem III.2.10 to see that $t^{-r}Z(t)\left(\frac{dx}{dt}\right)^2$ has at most a first order pole. Moreover $\Lambda_{t^{-r}}$ will be of the form $\dfrac{d}{dt^2} + \left(\dfrac{r-s+1}{t} + \text{holomorphic}\right)\dfrac{d}{dt}$ $+ \dfrac{1}{t}$ (holomorphic). It follows that $t^{-r}Z'(t)$ must be holomrophic so that

$$Z'(t) = t^r \text{ (holomorphic).}$$

To complete the proof we need only show that $\displaystyle\sum_{x \in X} r_x > 0$ where r_x is the lesser exponent of Λ at $x \in X$, as this would force Z' to have more zeroes than poles.

Now $\Lambda = \Lambda_{(J,\lambda)}$ where $J \in K(X)$ is non-constant and $\lambda^2 \in K(X)$ is non-zero. The effect of λ is to multiply the solutions of $\Lambda_{(J,1)}$ by λ which will not change the sum of the lower exponents over all of X. So without loss of generality we can assume $\lambda = 1$ and work with $\Lambda_{(J,1)}$. As our local calculations in Part II, §2 show, the lower exponents will be zero except where $J = 1$ or 0. If J has a zero of order k the lower exponent will be $\dfrac{-k}{6}$. The sum over all points where $J = 0$ will then give $\dfrac{-n}{6}$ where n is the valence of J (i.e. the degree of $X \xrightarrow{J} \mathbb{P}^1_{\mathbb{C}}$). Likewise the points where $J = 1$ will provide a sum of $\dfrac{n}{4}$. Thus the sum of the lower exponents will be $\dfrac{n}{12} > 0$. This completes the proof. □

Note that the sum of the lower exponents of Λ is independent of the Λ chosen to represent the Picard-Fuchs equation of E/X. It is always equal to 1/12 valence (or degree) of J. Again our proof actually shows more, namely:

Corollary III.5.8: No element of $T_3^{para}(\mathcal{Ol})$ is exact, or in other words, no element $Z \in L_\Lambda^{para}(\mathcal{Ol})$ is of the form $\Lambda Z'$. □

This shows that in Theorem III.4.1 it was not necessary to quotient out by $\Lambda K(X) \cap L_\Lambda^{para}(\mathcal{Ol})$, since this space is in fact zero, and we get that the rank r of the Mordell-Weil group $E^{gen}(K(X))$ of K(X)-rational points of the generic fiber of E/X is less than or equal to the dimension of

$L_\Lambda^{para}(\mathcal{O}l)$.

We now state the generalization of Shioda's result first due to Hoyt:

__Theorem III.5.9:__ Let E be a basic elliptic surface over a curve X having only multiplicative reduction (i.e. fiber types I_b $b \geq 1$). The space of generalized automorphic forms of weight 3 and of the first kind with respect to $\overline{\Gamma}$ is canonically isomorphic (over $\mathbb{C}$) to the space of holomorphic two forms on E, i.e.

$$S_3^1(\overline{\Gamma}) \cong H^0(E, \Omega_E^2).$$

__Proof:__ Let $\pi: E \to X$ be our projection and let $S \subset X$ denote as always the support of the singular fibers. Consider $E_0 = \pi^{-1}(X_0)$. The universal cover of E_0 is $\mathcal{h} \times \mathbb{C}$, $\mathcal{h}$ being the complex upper-half-plane and $\mathbb{C}$ the universal cover of any good fiber.

In Kodaira [16] a group action is defined on $\mathcal{h} \times \mathbb{C}$ so that the quotient is E_0 as follows:

For $\gamma \in \pi_1(X_0)$ and n_1, $n_2 \in \mathbb{Z}$ define an automorphism (γ, n_1, n_2) of $\mathcal{h} \times \mathbb{C}$ by

$$\mathcal{h} \times \mathbb{C} \xrightarrow{\ (\gamma, n_1, n_2)\ } \mathcal{h} \times \mathbb{C}$$

$$(z, t) \longrightarrow [\gamma z, (c_\gamma \omega(z) + d_\gamma)^{-1} (t + n_1 \omega(z) + n_2)]$$

where $\gamma \in \overline{\Gamma} \subset PSL_2(\mathbb{R})$, $\overline{\Gamma}$ being identified with $\pi_1(X_0)$; $\begin{pmatrix} a_\gamma & b_\gamma \\ c_\gamma & d_\gamma \end{pmatrix}$ is the monodromy; and ω is the period map. Now let $g \in S_3^1(\overline{\Gamma})$ be any form of the first kind and consider

$$\eta_g = g(z) \frac{d\omega(z)}{dz} \, dz \wedge dt$$

as a meromorphic 2-form on $\mathcal{h} \times \mathbb{C}$.

We show first that η_g is holomorphic. The only trouble can come at a point where $g(z)$ has a pole. The points $z \in \mathcal{h}$ where $g(z)$ has a pole are among the points lying over the cosingular points of Λ in X_0. If the exponent difference at such a point is $k \geq 1$ then in terms of a good local parameter, z for example, we have

$$g(z) = \frac{a_{-k+1}}{z^{k-1}} +$$

$$\frac{d\omega(z)}{dz} = c_{k-1} z^{k-1} + \dots \quad c_{k-1} \neq 0$$

this last because $\frac{d\omega}{dz} = \frac{-W}{\omega_2(z)^2}$ where W is the Wronskian computed with respect to z. Thus $g(z) \frac{d\omega(z)}{dz}$ is actually holomorphic and so is η_g.

Next we show that η_g is invariant under the action described above. This is just a computation:

$$g(\gamma z) \frac{d\omega(\gamma z)}{d(\gamma z)} d(\gamma z) \wedge d\left((c_\gamma \omega(z) + d_\gamma)^{-1} (t + n_1 \omega(z) + n_2) \right)$$

$$= (c_\gamma \omega(z) + d_\gamma)^3 g(z) \frac{d(M_\gamma \omega(z))}{dz} dz \wedge (c_\gamma \omega(z) + d_\gamma)^{-1} dt$$

$$= (c_\gamma \omega(z) + d_\gamma)^3 g(z) (c_\gamma \omega(z) + d_\gamma)^{-2} \frac{d\omega(z)}{dz} \wedge (c_\gamma \omega(z) + d_\gamma)^{-1} dt$$

$$= g(z) \frac{d\omega(z)}{dz} dz \wedge dt.$$

Thus η_g is a holomorphic invariant 2-form on $\mathcal{h} \times \mathbb{C}$ and so it descends to give a holomorphic two form on E_0, which we also denote η_g.

Finally, we want to show that η_g extends to a holomorphic two form on all of E. Let $x \in S$ be a point where the fiber E_x is of type I_b $b > 0$. Without loss of generality we can assume $\overline{\Gamma}$ has "$i\infty$" as a cusp, that this cusp lies over x, and that ω carries "$i\infty$" to "$i\infty$". In this way, we can choose

$$q = e^{2\pi i z/h} \qquad h > 0$$

to be a local coordinate in a neighborhood of $x \in S$ where $z \to z + h$ generates the isotropy group of $\overline{\Gamma}$ at "$i\infty$", and we will have that the monodromy matrix corresponding to going once around x in a counterclockwise manner is $\begin{pmatrix} 1 & b \\ 0 & 1 \end{pmatrix}$. Thus $\omega(q) = \frac{b}{2\pi i} \log q$ + holomorphic (see Part II, §2) and

$$g(z) \frac{d\omega(z)}{dz} dz \wedge dt = g(q) \frac{d\omega(q)}{dq} dq \wedge dt \quad ;$$

but $g \in S_3^1(\overline{\Gamma})$ which means that $g(q) = \sum_{n=1}^{\infty} a_n q^n = a_1 q + \dots$. It follows that in terms of q:

$$g(z) \frac{d\omega(z)}{dz} dz \wedge dt = (a_1 q + \dots) \cdot (\frac{b}{2\pi i} \frac{1}{q} + \dots) dq \wedge dt$$

$$= (\frac{a_1 b}{2\pi i} + \text{higher order terms}) dq \wedge dt$$

and so is holomorphic in q and t. Therefore, by the structure of a neighborhood of the singular fiber E_x (Kodaira [16]) our form η_g is holomorphic at every point of $B_x = E_x - \{b \text{ points}\}$. Hence η_g must be holomorphic in a neighborhood of E_x in E, and so, as we have seen, holomorphic on E.

This then gives us an injection

$$S_3^1(\overline{\Gamma}) \longrightarrow H^0(E, \Omega_E^2)$$

$$g \longrightarrow \eta_g \quad .$$

To see that it is also a surjection, note that any holomorphic two form η on E when restricted to E_0 gives an invariant two form on $\mathfrak{h} \times \mathbb{C}$ which must be of the form

$$h(z,t) \ dz \wedge dt$$

with h holomorphic in z and t. Fixing z and looking at the action by things of the form $(1,n_1,n_2)$, in other words the action in the fiber $z \times \mathbb{C} \subset \mathfrak{h} \times \mathbb{C}$, we see that

$$h(z, \ t + n_1 w(z) + n_2) = h(z,t)$$

must be a holomorphic elliptic function in t and therefore constant in t. Thus h is independent of t and we may write our invariant 2-form as

$$h(z) \ dz \wedge dt \ .$$

Setting $g(z) = h(z)\left(\dfrac{d\omega(z)}{dz}\right)^{-1}$, it is easy to check that $g(z) \ \epsilon \ S_3^1(\overline{\Gamma})$. (The condition at the cusps must hold since

$$g(z) \ \frac{d\omega(z)}{dz} \ dz \wedge dt$$

does extend to all of E.)

   This completes the proof.   □

   In Part IV we will extend this discussion to differentials of the second kind in relationship to the automorphic forms of the second kind.

   The theorem above leads to a number of corollaries. First observe that if $x \ \epsilon \ X$ and

$$g = \frac{\omega_2^3 \ z}{w^2}$$

is in $S_3^1(\overline{\Gamma})$ then

$$\text{ord}_x(Z(dx)^2) \geq s_x - 1$$

where s_x is the upper exponent of Λ at $x \in X$.

If we let $\mathcal{D}$ denote the divisor on X with

$$\text{ord}_x \mathcal{D} = -s_x + 1$$

then

$$\text{div}(Z) + \text{div}(dx)^2 \geq -\mathcal{D}$$

or

$$\text{div}(Z) + (\mathcal{D} + \text{div}(dx)^2) \geq 0 .$$

Thus

$$Z \in L \, (\mathcal{D} + \text{div}(dx)^2).$$

Conversely any such Z produces a generalized automorphic form of the first kind. Note that $Z \in L(\mathcal{D} + \text{div}(dx)^2)$ necessarily satisfies the parabolic residue conditions so that $L(\mathcal{D} + \text{div}(dx)^2) \subset L_\Lambda^{\text{para}}$.

We now define a very important divisor on X:

<u>Definition III.5.10</u>: We denote by $\mathcal{O}_0$ the divisor on X defined by

$$\mathcal{O}_0 = \mathcal{D} + \text{div}(dx)^2$$

where $\mathcal{D}$ is the divisor above given by

$$\mathcal{D} = \sum_{x \in X} (-s_x + 1)x,$$

s_x being the larger exponent of Λ at $x \in X$. □

<u>Corollary III.5.11</u>: $S_3^1(\tilde{\Gamma}) \cong H^0(E, \Omega_E^2) \cong L(\mathcal{O}_0) \subset L(\mathcal{O})$. □

Note also that $S_3^1(\overline{\Gamma})$ can be viewed as a subspace of $\dfrac{\text{second kind}}{\text{exact}}$ by our results above and that this latter space injects via parabolic cohomology into the period space (in fact it is isomorphic to it) which has dimension $4g - 4 + (\#S)$. Thus $p_g \leq 4g - 4 + (S\#)$. Later when we establish a Hodge structure on $\dfrac{\text{second kind}}{\text{exact}}$, we will actually get $2p_g \leq 4g - 4 + (\#S)$. This is in a sense the best possible estimate since for elliptic modular surfaces $p_g = 2g - 2 + \dfrac{(\#S)}{2}$ (see Shioda [27]).

PART IV.   HODGE THEORY

§1. The filtrations.

In this section we shall relate the various linear systems on the base curve X and the spaces of generalized automorphic forms that have been defined above to the Hodge theory of E over X.   In many case we will only state the relevant results.   Additional detail can be found in Stiller [32] and Cox and Zucker [4].

As always we wish to consider a basic elliptic surface E over X having only multiplicative reduction, i.e. singular fibers of type I_b b > 0.   Fixing a model for the generic fiber of E/X we calculated the differential equation Λ which represents the Picard-Fuchs equation for E/X, and after fixing a K-basis for Λ corresponding to E, we defined a number of linear systems and/or spaces of functions on X that were of interest to us.   Let's recall these spaces and the relationships between them.

We begin with the subspace $L_\Lambda^{para} \subset K(X)$ consisting of all functions Z which satisfy the parabolic residue conditions of Part II, §3.   This space corresponds to the space of generalized automorphic forms of the second kind having weight three which we have denoted T_3^{para}.   This correspondence is via the integral transforms of Part I, §3, but amounts to the map

$$L_\Lambda^{para} \xleftrightarrow{\;\widetilde{=}\;} T_3^{para}$$

$$Z \longleftrightarrow g = \frac{\omega_2^3 Z}{W} \;.$$

Next we have a number of subspaces of L_Λ^{para} and T_3^{para}.   First, the subspaces $L_\Lambda^{para}(\mathcal{O}\!\ell)$ and $T_3^{para}(\mathcal{O}\!\ell)$ where $\mathcal{O}\!\ell$ is the divisor $\mathcal{B} + \operatorname{div}(dx)^2$ defined in Theorem III.2.10.   We have $L_\Lambda^{para}(\mathcal{O}\!\ell) = L_\Lambda^{para} \cap L(\mathcal{O}\!\ell)$ where

$L(\mathcal{O}\!\!\mathit{l}) = \{Z \in K(X) \text{ s.t. } \operatorname{div} Z + \mathcal{O}\!\!\mathit{l} \geq 0\}$. Thus $L_{\Lambda}^{para}(\mathcal{O}\!\!\mathit{l})$ consists of those functions in the linear system $L(\mathcal{O}\!\!\mathit{l})$ which satisfy the parabolic residue conditions. $T_3^{para}(\mathcal{O}\!\!\mathit{l}) \subset T_3^{para}$ is defined by the conditions in Theorem III.3.1 and Theorem III.3.2, together with a set of linear conditions which amount to the residue conditions. Second we have $L_{\Lambda}^{para}(\mathcal{O}\!\!\mathit{l}_0)$ and $S_3^1(\overline{\Gamma})$. Here $\mathcal{O}\!\!\mathit{l}_0$ is the divisor $\mathcal{D} + \operatorname{div}(dx)^2$ defined in Corollary III.5.10. Note that $L_{\Lambda}^{para}(\mathcal{O}\!\!\mathit{l}_0) = L(\mathcal{O}\!\!\mathit{l}_0)$ because every function in $L(\mathcal{O}\!\!\mathit{l}_0)$ automatically satisfies the residue conditions. $S_3^1(\overline{\Gamma})$ are the generalized automorphic forms of the first kind which, as we have seen, are in one-to-one correspondence with the holomorphic two forms on E. Lastly, we have the exact elements $\{\Lambda Z', \; Z' \in K(X)\} \subset L_{\Lambda}^{para}$ and the exact generalized automorphic forms of the second kind and weight three which we denote by E_3.

We can summarize the relationships that exist between these spaces by the following diagram:

We also have

$$L_\Lambda^{para}(\mathcal{O}_0) \subset L_\Lambda^{para}(\mathcal{O}) \hookrightarrow \frac{L_\Lambda^{para}}{\{\Lambda Z'\}}$$

$$S_3^1(\overline{T}) \subset T_3^{para}(\mathcal{O}) \hookrightarrow \frac{T_3^{para}}{E_3} = \frac{\text{second kind}}{\text{exact}} .$$

The last two maps being injections by the results in Corollary III.5.6, Theorem III.5.7, and Corollary III.5.8. It is worth pointing out that L_Λ^{para}, $L_\Lambda^{para}(\mathcal{O})$, $L_\Lambda^{para}(\mathcal{O}_0)$ and $\{\Lambda Z'\}$ depend on a choice of model for the generic fiber of E/X and on the choice of a parameter x used in forming Λ, whereas the various spaces of automorphic forms do not.

We would like to show that the above filtrations are functorial in a suitable sense. To do this we consider first a special case. The proofs in the general case are identical. We let E/X denote a basic elliptic surface having only multiplicative reduction, i.e. singular fiber types I_b $b > 0$ only. After fixing a model for the generic fiber, we obtain a K-equation Λ and can choose a K-basis ω_1, ω_2 corresponding to E/X as usual. Fixing a base point $x_0 \in X_0 = X-S$, where S is the set of true singular points, we obtain the monodromy representation $\pi_1(X_0, x_0) \to SL_2(\mathbb{Z})$ via the analytic continutation of ω_1, ω_2. We denote the global monodromy group in $SL_2(\mathbb{Z})$ by M where we assume for the moment that $-I_2$ is not in M, and we denote its projectivization in

$PSL_2(\mathbb{Z})$ by $\overline{M}$. Recall that E can be regarded as the pull-back of the canonical elliptic modular surface E_M over the modular curve $\mathring{\hbar}^*/\overline{M}$ via the map $X \xrightarrow{[\omega]} \mathring{\hbar}^*/\overline{M}$ induced by the period map $\omega = \omega_1/\omega_2$:

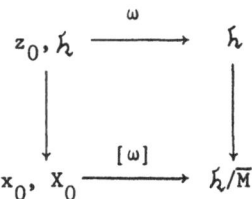

Now if $g \in S_3(M)$ is a cusp form of weight three for $M \subset SL_2(\mathbb{Z})$ in the classical sense then it is easy to see that $g \circ \omega$ is in $S_3^1(\overline{\Gamma})$. In fact considerably more is true.

<u>Theorem IV.1.1</u>: Let g be any automorphic form of weight three for M which is also of the second kind, then $g \circ \omega$ is of the second kind for $\overline{\Gamma}$ ($\overline{\Gamma} \subset PSL_2(\mathbb{R})$) being such that $\mathring{\hbar}/\overline{\Gamma} \cong X_0$ etc.). Thus $g \circ \omega \in T_3^{para}$ for E/X.

<u>Proof</u>: For simplicity we shall assume M is such that E_M has only fibers of type I_b $b > 0$. (Thus M is torsion free and has only regular cusps.) We choose a K-equation Λ^M to represent the Picard-Fuchs equation of E_M over $\mathring{\hbar}^*/\overline{M}$. There is, up to scalar multiples, only one K-basis ω_1^M, ω_2^M of Λ^M corresponding to E_M over $\mathring{\hbar}^*/\overline{M}$ with monodromy representation equal to the identity. (Here we identify $\pi_1(\mathring{\hbar}/\overline{M})$ with M in the obvious way.) We denote by $\Lambda^M \circ [\omega]$ the pull-back of Λ^M to a differential equation on $X_0 \cong \mathring{\hbar}/\overline{\Gamma}$. It will be a K-equation with K-basis $\omega_1^M \circ [\omega]$, $\omega_2^M \circ [\omega]$, and $\Lambda^M \circ [\omega]$ will represent the Picard-Fuchs equation of E over X. It will thus differ

from the original choice of Λ, ω_1, ω_2 used to form our period map by a twist. That is there will exist some function $h \in K(X)$ so that

$$\Lambda_h = \Lambda^M \circ [\omega]$$

$$h\omega_1 = \omega_1^M \circ [\omega]$$

$$h\omega_2 = \omega_2^M \circ [\omega].$$

Note that $\Lambda_{ch} = \Lambda_h$ for $c \in \mathbb{C}^*$, so we can always account for the fact that ω_i^M are only determined up to scalars.

As the notion of second kind is independent of twist we can assume without any loss of generality that $\Lambda = \Lambda^M \circ [\omega]$ and $\omega_i = \omega_i^M \circ [\omega]$ on X.

Now suppose that g is of the second kind for M, so that g can be written

$$\frac{(\omega_2^M)^3 z^M}{(w^M)^2}$$

where w^M is the Wronskian of Λ^M taken with respect to some parameter x^M and z^M is a modular function in $L_{\Lambda^M}^{para}$. (Note that ω_2^M is a (possibly meromorphic) automorphic form of weight one for $M \subset SL_2(\mathbb{Z})$.) We have that on h

$$g \circ \omega = \frac{(\omega_2^M \circ \omega)^3 (z^M \circ \omega)}{(w^M \circ \omega)^2} = \frac{\omega_2^3 z}{w^2} \qquad \begin{array}{l} Z = z^M \circ [\omega] \\[2mm] W = w^M \circ [\omega] \end{array}$$

with $Z \in K(X)$ and W equal to the Wronskian of Λ <u>with respect to the parameter</u> $x^M \circ [\omega]$ on X. To check that $g \circ \omega$ is of the second kind, we must see that $Z = z^M \circ [\omega]$ is in L_Λ^{para}. This involves looking at the residues of

$$\frac{-\omega_2 Z}{W} d[x^M \circ \omega] \qquad\qquad \frac{\omega_1 Z}{W} d[x^M \circ \omega]$$

and

$$\frac{\tilde{\omega}Z}{W} d[x^M \circ \omega] \qquad\qquad \tilde{\omega} \text{ invariant}$$

at the cosingular and singular points respectively. In all cases the differentials are the pull-backs via $[\omega]$ of the corresponding ones for Z^M on $\mathfrak{z}^*/\overline{M}$ which have no residue by assumption. It follows that they have no residue either. □

Corollary IV.1.2: g is exact if and only if $g \circ \omega$ is exact.

Proof: If we express Λ in terms of the parameter $x^M \circ [\omega]$ the result is clear since we would have $Z = Z^M \circ [\omega] = (\Lambda^M Z^{,M}) \circ [\omega] = \Lambda(Z^{,M} \circ [\omega]) = \Lambda Z'$. The converse follows because Λ is a pullback and K-equations have no algebraic solutions. □

Note that if x^M is a good local parameter at a point $p \in \mathfrak{z}^*/\overline{M}$ then $x^M \circ [\omega]$ need not be a good local parameter at a point $x \in X$ which lies over p (unfortunately the notation here is bad). good local parameter at $x \in X$ (e.g. $t = q = e^{2\pi i z/h}$ at the cusps of $\overline{\Gamma}$) then in terms of this good parameter

$$g \circ \omega = \frac{\omega_2^3 Z \left(\frac{d(x^M \circ \omega)}{dt}\right)^2}{W^2 \left(\frac{d(x^M \circ \omega)}{dt}\right)^2}$$

where $Z\left(\frac{d(x^M \circ \omega)}{dt}\right)^2 \in L_{\Lambda^{new}}^{para}$ with Λ^{new} equal to Λ expressed in terms of t and

$$W^{new} = W \frac{d(x^M \circ \omega)}{dt}$$

being the Wronskian of Λ^{new} computed with respect to t. (These parameter change effects are also discussed in Part I, §2.)

We can generalize the above results to the following case: Let E be any basic elliptic surface over X having only multiplicative reduction, i.e. fibers of type I_b $b > 0$, and let X' $\xrightarrow{f}$ X be any other

smooth curve X' together with a map f onto X. Consider the basic

elliptic surface E' over X' induced by pull-back via f. (Note E'

has only fibers of type I_b $b > 0$.) We have:

Theorem IV.1.3: f induces an injective map f* from second kind modulo

exact for E over X to second kind modulo exact for E' over X'.

Proof: This is clear if as above we represent E' over X' via the

pull-back of Λ for E over X with the pulled-back K-basis and with

respect to the pulled-back parameter.   □

A stronger result can be derived. Consider the filtration $F_E^{\bullet}$ for E/X

$$F_E^{\bullet}: \quad F_E^0 \supset F_E^1 \supset F_E^2 \supset 0$$

where

$$F_E^0 = \frac{\text{second kind}}{\text{exact}} = \frac{T_3^{\text{para}}}{E_3}$$

$$F_E^1 = \text{image of } T_3^{\text{para}}(\mathcal{M}) \text{ injected into } F_E^0$$

$$F_E^2 = \text{image of } S_3^1(\overline{T}) \text{ injected into } F_E^0$$

or the corresponding filtration on $L_\Lambda^{\text{para}}/\{\Lambda Z'\}$. We then have:

Theorem IV.1.4: Let E' over X' be the basic elliptic surface induced

by E over X via f: X' $\rightarrow$ X, then f induces an injection

f*: $F_E^{\bullet} \rightarrow F_{E'}^{\bullet}$ which preserves filtration level. Moreover the induced map

of associated graded objects

$$\mathrm{gr}(F_E^{\:\bullet}) \to \mathrm{gr}(F_{E'}^{\:\bullet})$$

is an injection.

Proof: As E' is induced by the map $f: X' \to X$ from E over X, it is easy to see that E' will have singular fibers also only of type I_b $b > 0$, and that the support S' of these singular fibers in X' will be $f^{-1}(S)$. We get a commutative diagram

$$
\begin{array}{ccc}
\mathcal{h} & \xrightarrow{\;f\;} & \mathcal{h} \\
\downarrow & & \downarrow \\
X_0' & \xrightarrow[\;f\;]{} & X_0
\end{array}
$$

where $X_0' = X' - S'$ and where the map $\mathcal{h} \to \mathcal{h}$ is induced by f (we denote it by f as well).

As we have already seen, the generalized automorphic forms of the second kind for E pull-back to forms of the second kind for E' with only exact forms pulling back to exact forms. Thus f induces an injection $f^*: F_E^0 \to F_{E'}^0$.

Now suppose $g \in T_3^{\mathrm{para}}(\mathcal{O}\!\mathcal{L}) \overset{\sim}{=} F_E^1$. So g is a generalized automorphic form for E satisfying the required local conditions that it be holomorphic and vanishing at the cusps (points of S) and that elsewhere it have order at worst $-2k+1$ where k is the exponent difference of Λ, the differential equation for E over X. Note that the residue conditions are automatic. We must see that $g \circ f$ is in $T_3^{\mathrm{para}}(\mathcal{O}\!\mathcal{L}')$ where $\mathcal{O}\!\mathcal{L}'$ is determined by the pulled-back equation Λ' with respect to the pulled-back parameter. Clearly $g \circ f$ will be holomorphic and vanishing at the cusps (= S'), so it only remains to check the behavior at the cosingular points. Let $x' \in X_0'$ with $f(x') = x \in X_0$ and suppose the exponent difference for Λ at x is k and that the ramification at x' over x is e, then the exponent difference k' for Λ' at x' is easily seen to be ke

$$k' = ke.$$

It follows that g $\circ$ f will have order at worst

$$(-2k+1)e$$

and

$$(-2k+1)e = -2ke + e \geq -2ke + 1 = -2k' + 1.$$

Thus g $\circ$ f is in $T_3^{para}(\mathcal{O}\mathcal{l}')$.

If g ϵ T_3^{para} is arbitrary but g $\circ$ f ϵ $T_3^{para}(\mathcal{O}\mathcal{l}')$ then g must be holomorphic and vanishing at the cusps (= S $\subset$ X) and if g has order r at a point x ϵ X_0 then we must have

$$re \geq -2ke + 1$$

so

$$r \geq -2k + \frac{1}{e}$$

which forces

$$r \geq -2k + 1.$$

Thus g must have been in $T_3^{para}(\mathcal{O}\mathcal{l})$. We conclude that there are induced maps

$$f*: F_E^1 \rightarrow F_{E'}^1$$

$$f*: F_E^0/F_E^1 \rightarrow F_{E'}^0/F_{E'}^1$$

and that these maps are injective. The proof for g of the first kind is similar and the result follows. $\qquad \square$

Recall that in Part II, §3 we associated to each generalized automorphic form in T_3^{para} (second kind) a parabolic cohomology class by taking periods and that the exact elements, $E_3 \subset T_3^{para}$, were precisely the ones that gave the trivial parabolic cohomology class. Thus we had an injection

$$F_E^0 \cong \frac{\text{second kind}}{\text{exact}} \cong \frac{T_3^{para}}{E_3} \xrightarrow{\quad [\ ]\quad} H^{para}(\overline{\Gamma}, \rho)$$

$$g \longmapsto [g]$$

where $H^{para}(\overline{\Gamma}, \rho)$ denotes the parabolic cohomology and $\rho: \overline{\Gamma} \to SL_2(\mathbb{Z})$ is the monodromy representation for our fixed choices of Λ, ω_1, ω_2.

<u>Theorem IV.1.5:</u> The map from $\dfrac{\text{second kind}}{\text{exact}} = \dfrac{T_3^{para}}{E_3} = F_E^0$ to H^{para} is an isomorphism.

<u>Proof:</u> See Stiller [32]. □

Via this isomorphism we can give the complex vector space T_3^{para}/E_3 an underlying real structure by looking at those elements whose periods lie in $\mathbb{R}$ and an underlying lattice by taking those with integral, $\mathbb{Z}$, periods. One can in fact interpret our filtration as a Hodge filtration with associated Hodge structure of weight 2. Thus

$$F_E^0 \cong H^{2,0} \oplus H^{1,1} \oplus H^{0,2} .$$

We shall see what this amounts to geometrically in a moment. First recall that the space $F_E^2 \cong S_3^1(\overline{\Gamma})$ of generalized automorphic forms of the first kind and weight three is canonically isomorphic to $H^0(E, \Omega_E^2)$. Thus our $H^{2,0}$ above is naturally identified with $H^0(E, \Omega_E^2)$.

In Part III, §2 we discussed the exact sequence of sheaves of abelian groups on X

$$0 \to G \to \mathcal{O}(\mathcal{f}) \to \Omega(B_0) \to 0$$

where G is the homological invariant $\tilde{=} R^1\pi_*(Z)$ and $\pi: E \to X$ is the projection. This leads to the exact sequence

1) $\qquad 0 \to H^0(X, \Omega(B_0)) \longrightarrow H^1(X,G) \overset{i}{\longrightarrow} H^1(X, \mathcal{O}(\mathcal{f})) \to \quad .$

__Theorem IV.1.6__: $H^1(X,G) \otimes_{\mathbb{Z}} \mathbb{C} \tilde{=} H^1(X, R^1\pi_*(\mathbb{C}))$ is isomorphic to H^{para} in a natural way.

__Proof__: See Shioda [27]. $\qquad\qquad\qquad\qquad\qquad\qquad\qquad\qquad\qquad$ □

__Corollary IV.1.7__: The map $H^1(X,G) \otimes_{\mathbb{Z}} \mathbb{C} \to H^1(X, \mathcal{O}(\mathcal{f}))$ induced by i in 1) above corresponds to the map $F_E^0 \to F_E^0/F_E^1 \tilde{=} H^{0,2}$.

__Proof__: It suffices to note that Kodaira (Kodaira [17]) identifies the space $H^0(X, \mathcal{O}(K - \mathcal{f}))$ with $H^0(E, \Omega_E^2)$ where K is the canonical divisor on X. Serre duality then gives the result. (See also Shioda [27] for the specific isomorphism $H^1(X, \mathcal{O}(\mathcal{f})) \tilde{=} H^2(E, \mathcal{O}_E)$.) $\qquad\qquad$ □

Note that the sections of E/X $(\tilde{=} H^0(X, \Omega(B_0)))$ after $\otimes_{\mathbb{Z}} \mathbb{Q}$ are in the kernel of this map which agrees with the fact that to each section we have associated an element of F_E^1 $(\tilde{=} T_3^{para}(\mathcal{O}\mathcal{l})$ or $L_\Lambda^{para}(\mathcal{O}\mathcal{l}))$. In fact sections give rise to elements in $H^{1,1}$, i.e. of type (1,1), as we should expect. We also remark that Shioda has shown that over $\mathbb{R}$ the image under i of $H^1(X,G)$ in $H^1(X, \mathcal{O}(\mathcal{f}))$ spans the entire space which has real dimension $2p_g$. This again is what we would expect since the kernel of the projection from $H_{\mathbb{R}}^{para}$, the real parabolic cohomology, onto $H^{0,2}$ has outside its kernel $(H^{2,0} \oplus H^{0,2})_{\mathbb{R}}$, i.e. the real classes in $H^{2,0} \oplus H^{0,2}$, which has real dimension $2p_g$.

In order to understand geometrically the source of this filtration, we consider the Leray spectral sequence for $\pi: E \to X$ and the constant sheaf $\mathbb{C}$ on E:

$$E_2^{p,q} = H^p(X, R^q \pi_*(\mathbb{C}))$$

$$E_\infty \Rightarrow H^*(E, \mathbb{C}) \quad .$$

Now $E_2^{p,q} = 0$ except for $p = 0,1,2$ and $q = 0,1,2$. Moreover $R^1 \pi_*(\mathbb{Z}) = G$ the homological invariant of E/X, and it is well-known that

$$H^0(X, G) = 0$$

and that

$$H^2(X, G) = \text{finite group}.$$

Thus the Leray spectral sequence above degenerates at E_2, and the Leray filtration on $H^2(E, \mathbb{C})$ is $L^2 \subseteq L^1 \subseteq L^0 = H^2(E, \mathbb{C})$ where

$$L^1 = \ker(H^2(E, \mathbb{C}) \to H^0(X, R^2 \pi_*(\mathbb{C})),$$

so that L^1 consists of those classes in $H^2(E, \mathbb{C})$ which restrict to zero on the fibers of $\pi: E \to X$,

$$L^1/L^2 = H^1(X, R^1 \pi_* (\mathbb{C})),$$

and $L^2 = \text{im}(H^2(X, \mathbb{C}) \to H^2(E, \mathbb{C})) = \mathbb{C}[E_x] \quad (x \in X)$, so that L^2 consists of multiples of the fiber.

As we have seen $L^1/L^2 \cong H^1(X, R^1 \pi_*(\mathbb{C}))$ is isomorphic to the parabolic cohomology group $H^{para} \cong F_E^0$.

Theorem IV.1.8: (Cox and Zucker [4]) There is a natural Hodge structure of weight two on $H^1(X, R^1 \pi_*(\mathbb{C}))$ and this Hodge structure corresponds to the one induced by the Hodge structure on $H^2(E, \mathbb{C})$ through the Leray spectral sequence. □

This theorem explains the source of the filtration $F_E^{\cdot}$. For further details see Cox and Zucker [4], Shioda [27], Hoyt [12] and [13], and Stiller [36]. In particular the above Hodge structure has a polarization and the pairing is discussed geometrically in Cox and Zucker [4] and in terms of the generalized automorphic forms in Hoyt [13].

§2. Differentials of the second kind.

In this section we shall investigate the connection between L_Λ^{para}, T_3^{para} and differentials of the second kind on E. Recall that a meromorphic 2-form ϕ on E is said to be a differential of the second kind if it is closed and if there exists a divisor D on E with complement $U = E - D$ such that ϕ is holomorphic on U and is in the image of

$$H^2(E,\mathbb{C}) \to H^2(U,\mathbb{C}).$$

As usual we let $S \subset X$ be the set of true singularities, so that for $s \in S$ the fiber E_s of E over X will be of type I_{b_s} (we still are assuming multiplicative reduction). We denote the complement of S in X by X_0, as usual, and by E_0 the Zariski open set $\pi^{-1}(X_0)$ in E. Recall that E_0 can be constructed by taking $\hbar \times \mathbb{C}$, its universal cover, and considering the group action

$$\hbar \times \mathbb{C} \xrightarrow{\quad (\gamma,n_1,n_2) \quad} \hbar \times \mathbb{C}$$

$$(z,t) \longmapsto (\gamma z, (c_\gamma \omega(z)+d_\gamma)^{-1}(t+n_1\omega(z)+n_2))$$

where $\gamma \in \bar{\Gamma} \subset PSL_2(\mathbb{R})$ is identified with $\gamma \in \pi_1(X_0)$ via the universal cover $\hbar \to X_0 \cong \hbar/\bar{\Gamma}$; $\begin{pmatrix} a_\gamma & b_\gamma \\ c_\gamma & d_\gamma \end{pmatrix}$ is the monodromy; $n_1, n_2 \in \mathbb{Z}$; and

$\omega(z) = \omega_1(z)/\omega_2(z)$ is the period map (see Theorem III.5.9 and its proof for details).

Theorem IV.2.1: Let $g \in T_3^{para}$ (so $g = \dfrac{\omega_2^3 Z}{W^2}$ for $Z \in L_\Lambda^{para}$) be a generalized automorphic form of the second kind, then

$$\phi_g = g(z) \frac{d\omega(z)}{dz} \, dz \wedge dt = \frac{-\omega_2 Z}{W} \, dz \wedge dt$$

is a meromorphic 2-form on $h \times \mathbb{C}$ which is invariant under the above action and hence gives a meromorphic 2-form on E_0. This form extends to a meromorphic two form on E which will be a differential of the second kind. Thus we have an injective map

$$\begin{array}{c} T_3^{para} \\ \text{or} \\ L_\Lambda^{para} \end{array} \rightarrowtail \longrightarrow \text{2-forms of the second kind} .$$

Proof: As in Theorem III.5.9 we can easily determine that $\phi_g = $

$g(z) \dfrac{d\omega(z)}{dz} \, dz \wedge dt$ is a meromorphic 2-form on $h \times \mathbb{C}$ which is invariant under the group action described above. Thus ϕ_g descends to a meromorphic 2-form on E_0 which we also denote by ϕ_g. Moreover because $g(z)$ is meromorphic at the cusps the same argument shows that ϕ_g extends to all of E as a meromorphic 2-form. The only really new part in the theorem is to show that ϕ_g on E is a differential of the second kind.

To show this we must consider the residues. Let D be a divisor (which can be taken to be a linear combination of fibers of E/X) such that ϕ_g is holomorphic on $U = E - D$. We define a residue to be an integral

$$\int_\Delta \phi_g$$

where $\Delta \in H(U,\mathbf{Z})$ is a 2-cycle that is homologous to zero in E. It is easy to see that ϕ_g is of the second kind if and only if it has no residues in open sets $U = E - D$ for sufficiently large divisors D.

Note that on $\hbar \times \mathbb{C}$

$$\phi_g = g(z)\,\frac{d\omega(z)}{dz}\,dz \wedge dt$$

$$= \frac{\omega_2^3(z)Z(z)}{W^2(z)}\,\frac{-W}{\omega_2^2(z)}\,dz \wedge dt$$

$$= \frac{-\omega_2(z)Z(z)}{W(z)}\,dz \wedge dt.$$

The differential $\dfrac{-\omega_2(z)Z(z)}{W(z)}\,dz$ is of course one that arises in the parabolic residue conditions of Part II, §3.

As the divisor D on which ϕ_g has poles consists only of the fibers of $E \xrightarrow{\pi} X$, we consider a point $x \in X_0$ and in X_0 a small loop γ_x going once around x in a counterclockwise manner:

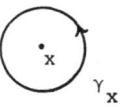

The family of elliptic curves given by E is locally topologically trivial over a small disc D_x about x and can be given by the lattices L_x generated by 1 and $\omega(x)$. Let τ stand for the cycle on the fiber gotten by going from 0 to 1 along the real axis, then $\gamma_x \times \tau$ is an element of $H_2(U,\mathbf{Z})$, for a suitable $U = \pi^{-1}(V)$ (V some Zariski open subset of X not containing x), which is homologous to zero in E, being the boundary of $D_x \times \tau$.

The residue

$$\int_{\gamma_x \times \tau} \phi_g$$

is

$$\int_0^1 \int_{\gamma_x} \frac{-\omega_2 Z}{W} \, dx dt = \int_{\gamma_x} \frac{-\omega_2 Z}{W} \, dx = 0$$

because $Z \in L_\Lambda^{para}$ or equivalently $g \in T_3^{para}$. Had we chosen a different cycle on the fiber the result would have been an integral of the form

$$\int_{\gamma_x} \frac{(a\omega_1 + b\omega_2)Z}{W} \, dx \qquad a, b \in \mathbb{Z}$$

which is still zero by the residue condition (see Definition II.3.3). Thus the residue condition of Part II, §3 gives us zero residues for ϕ_g in these instances.

At points $s \in S$ the family is no longer topologically trivial since the fiber at s is degenerate of type I_{b_s}. However we can construct a class Δ in $H_2(U, Z)$ homologous to zero in E in the same manner by using the invariant cycle at and near s. The resulting residue

$$\int_\Delta \phi_g$$

is

$$\int_{\gamma_x} \frac{\tilde{\omega} Z}{W} \, dx = 0$$

where $\tilde{\omega}$ is an invariant solution at $s \in S$, and this is zero by the parabolic residue condition.

It now follows easily that ϕ_g is of the second kind. It is exactly the parabolic residue conditions on $Z \in L_\Lambda^{para}$ that give this result! □

Recall that a differential of the second kind will be called <u>exact</u> if it is d of a meromorphic 1-form.

Consider the map

$$L_\Lambda^{para}(\theta\nu) \quad \rightarrow \quad \frac{\text{2-forms of the second kind}}{\text{d (meromorphic 1-forms)}}$$

$$Z \quad \rightarrow \quad \phi_g$$

which is defined as above.

<u>Theorem IV.2.2</u>: The kernel of this map is a compelx vector space of dimension r where r is the rank of the Mordell-Weil group of $E^{gen}/K(X)$.

<u>Proof</u>: It is well-known that differentials of the second kind modulo exact differentials is naturally isomorphic to

$$\frac{H^2(E,\mathbb{C})}{\mathbb{C}\text{-span of the algebraic classes}}$$

which has dimension $b_2 - \rho$ with b_2 the second betti number of E and ρ the Picard number. It is fairly easy to see that the kernel of our map consists of the $\mathbb{C}$-span of the $Z \in L_\Lambda^{para}(\theta\nu)$ associated to the sections and so has dimension r.

PART V.   THE PICARD NUMBER

§1.   Periods and period integrals.

In this section we shall outline a method for computing the Picard number of an elliptic surface and in the last section apply the method in a number of examples. In order to minimize the technical difficulties and to avoid lengthy computations we shall restrict ourselves to a specific class of elliptic surfaces. Many of the restrictions we shall impose are unnecessary and the reader who has managed to get this far will undoubtedly see what modifications are needed to handle other cases.

Let E denote a basis elliptic surface (see page 41 above) over a curve X. We shall assume that E has only multiplicative reduction, that is, singular fibers of type I_b, $b > 0$ only. In addition we shall assume E is a regular surface so that $q = \dim_{\mathbb{C}} H^1(E, \mathcal{O}_E) = 0$. In this case X will be isomorphic to $\mathbb{P}^1_{\mathbb{C}}$. This assumption is not necessary but makes things easier to write down.

We begin by choosing a model for the generic fiber of E over X which we denote by E^{gen} over K(X) the function field of X ($\cong \mathbb{P}^1_{\mathbb{C}}$):

1) $$Y^2 = 4X^3 - g_2 X - g_3 \qquad g_2, \ g_3 \in K(X) .$$

Using Theorem III.1.1 and Theorem II.1.3 we can explicitly calculate the K-equation $\Lambda = \Lambda_{(J,\lambda)}$ which annihilates the periods of $\frac{dX}{Y}$ for the curve given by 1) above. $\Lambda = \frac{d^2}{dx^2} + P \frac{d}{dx} + Q$ with $P, Q \in K(X)$ given in terms of J, λ. Here J is the functional invariant (i.e. J-invariant) of E over X and $\lambda^2 \in K(X)$. Corresponding to E is a K-basis ω_1, ω_2 (see

Definition II.1.1) of Λ which is unique up to the action of $SL_2(\mathbb{Z})$ and/or scalar matrices. In fact once the model 1) is fixed, we can normalize ω_1, ω_2 so that only the action of $SL_2(\mathbb{Z})$ remains by making the requirement that ω_1, ω_2 be exactly the periods of $\frac{dX}{Y}$.

Set S denote the finite set of points in X over which E has a singular fiber. For $s_i \in S$ the singular fiber will be of type I_{b_i}, $b_i > 0$ because of ou assumptions. The number b_i will be order of the pole of J at s_i. Using a formula of Shioda (see Introduction) we find that the Picard number ρ can be computed in terms of the rank r of the Mordell-Weil group, $E^{gen}(K(X))$, of K(X)-rational points on the generic fiber, E^{gen}, of E/X. We have:

$$\rho = r + 2 + \sum_{s_i \in S} (b_i - 1)$$

$$= r + 2 + \mu - \#S$$

where μ = valence J, i.e. the degree of the map $X \xrightarrow{J} \mathbb{P}^1_{\mathbb{C}}$.

In order to determine r, recall that there is an injection

$$E^{gen}(K(X))/\text{torsion} \hookrightarrow L^{para}_{\Lambda}(\mathcal{O}\!\ell) \subset L(\mathcal{O}\!\ell)$$

$$\wp \longrightarrow \Lambda \int_{\mathcal{O}}^{\wp} \frac{dX}{Y} = Z$$

where the divisor $\mathcal{O}\!\ell$ is easily computed once Λ is given, and that the image is exactly those $Z \in L^{para}_{\Lambda}(\mathcal{O}\!\ell)$ with integer periods (see Theorem III.3.4). Thus we must investigate how to comute the periods of $Z \in L^{para}_{\Lambda}(\mathcal{O}\!\ell)$.

Denote by X_0 the Zariski open set X-S and pick a base point $x_0 \in X_0$ which we will assume to be chosen away from the singularities of Λ in X_0 (the so-called cosingularities). Since $X \cong \mathbb{P}^1_{\mathbb{C}}$, X_0 is a punctured sphere and $\pi_1(X_0,x_0)$ is a free group on #S generators subject to one relation (#S $\geq$ 4 under our assumptions). If $S = \{s_1,\ldots,s_t\}$ we can choose generators γ_1 , $\ldots$, γ_t to be simple loops around s_1, $\ldots$, s_t respectively. The picture will be

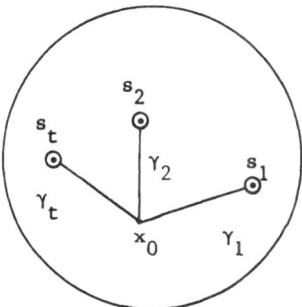

The one relation is $\gamma_1 \cdot \gamma_2 \cdot \ldots \cdot \gamma_t = 1$. In addition we can arrange that the γ_i do not pass through any cosingular point in X .

Now fix normalized branches of ω_1, ω_2 at x_0 once and for all. Given $Z \in L_\Lambda^{para}(\mathcal{O}\!\!\ell) \subset L_\Lambda^{para}$ we consider the inhomogeneous equation

$$\Lambda f = Z$$

which has for solutions near x_0 the functions

2) $$f(x) = (\int_{x_0}^x \frac{-\omega_2 Z}{W} \, dx + c_1)\omega_1 + (\int_{x_0}^x \frac{\omega_1 Z}{W} \, dx + c_2)\omega_2 \qquad c_1, \, c_2 \in \mathbb{C}$$

(see Theorem I.2.1). Because the residue conditions hold for Z, such a function f will continue as a meromorphic multivalued function on all of X_0. We fix a branch of f at x_0, i.e. choose specific c_1, c_2, and once this is done the periods $[m_\gamma, n_\gamma]$ for $\gamma \in \pi_1(X_0,x_0)$ are given by analytic (meromorphic) continuation around the path — f returning to another solution $f + m_\gamma \omega_1 + n_\gamma \omega_1$ of $\Lambda f = Z$ at x_0.

Recall that for any paths $\gamma, \tau \in \pi_1(X_0,x_0)$ we have

$$[m_{\gamma\tau}, \, n_{\gamma\tau}] = [m_\gamma, \, n_\gamma]M_\tau + [m_\tau, \, n_\tau]$$

where $\gamma\tau$ means first follow γ and then τ and where M is the monodromy matrix for the continuation of our fixed branches of ω_1, ω_2. Thus the periods are completely determined by their values on the generators γ_1, ..., γ_t with the one relation

3) $\quad [0,0] = [m_{\gamma_1}, n_{\gamma_1}] M_{\gamma_2} \ldots M_{\gamma_t} + \ldots + [m_{\gamma_i}, n_{\gamma_i}] M_{\gamma_{i+1}} \ldots M_{\gamma_t}$

$$+ \ldots + [m_{\gamma_t}, n_{\gamma_t}].$$

The assignment to any path $\gamma \in \pi_1(X_0, x_0)$ of the periods around γ gives rise to a parabolic cocycle for the monodromy representation (the usual condition "at the cusps" holds because $Z \in L_\Lambda^{para}(\theta_1) \subseteq L_\Lambda^{para}$). If we were to choose another solution to $\Lambda f = Z$, that is choose different c_1, c_2, in 2) above, then the periods are adjusted by a parabolic coboundary. Thus $Z \in L_\Lambda^{para}$ determmnes a well-defined parabolic cohomology class.

In all that we have done in the last few paragraphs branches of ω_1, ω_2 have been fixed at x_0. It will be important to see what happens when we change from ω_1, ω_2 to another normalized K-basis ω_1', ω_2' for E/X. We will have

$$\begin{pmatrix} \omega_1' \\ \omega_2' \end{pmatrix} = N \begin{pmatrix} \omega_1 \\ \omega_2 \end{pmatrix} \qquad N \in SL_2(\mathbb{Z}) \quad .$$

The new monodromy representation will be

$$\pi_1(X_0, x_0) \rightarrow SL_2(\mathbb{Z})$$

$$\gamma \rightarrow N M_\gamma N^{-1} = M_\gamma'$$

where M_γ is the monodromy matrix for ω_1, ω_2. Of course our chosen

solution of $\Lambda f = Z$ will be expressable as

$$f(x) = (\int_{x_0}^x \frac{-\omega_2' Z}{W} \, dx + c_1')\omega_1' + (\int_{x_0}^x \frac{\omega_1' Z}{W} \, dx + c_2')\omega_2' .$$

An easy calculation shows that the periods in terms of this new basis will

be

$$[m_\gamma', \ n_\gamma'] = [m_\gamma, \ n_\gamma]N^{-1} .$$

For each of our generators γ_i choose a matrix $N_i \ \epsilon \ SL_2(\mathbb{Z})$

(unique up to left multiplication by $\pm\begin{pmatrix} 1 & n \\ 0 & 1 \end{pmatrix}$ $n \ \epsilon \ \mathbb{Z}$) so that the K-basis

$$(\begin{matrix} \omega_{1i} \\ \omega_{2i} \end{matrix}) = N_i \ (\begin{matrix} \omega_1 \\ \omega_2 \end{matrix})$$

is normalized to have monodromy $\begin{pmatrix} 1 & b_i \\ 0 & 1 \end{pmatrix}$ $b_i \ \epsilon \ \mathbb{Z}$, $b_i > 0$ around γ_i

when the singular fiber is type I_{b_i} at s_i. This means that near s_i

the continuation of ω_{2i} along γ_i is a local invariant solution to

$\Lambda f = 0$. In particular,

$$M_i = N_i^{-1} \begin{pmatrix} 1 & b_i \\ 0 & 1 \end{pmatrix} N_i .$$

where M_i is the original monodromy in terms of ω_1, ω_2 around γ_i.

Our previous calculations show that in terms of the basis ω_{1i}, ω_{2i}

the periods around γ_i will be

$$[0, \ n_i] = [0, \ \chi_{2i}(\gamma_i)] + [0, \ b_i c_{1i}]$$

108

where $\chi_{1i}(\gamma_i) = 0$ because $-\omega_{2i}$ is invariant at s_i and

$$\chi_{1i}(\gamma_i) = \int_{\gamma_i} \frac{-\omega_{2i}Z}{W} \, dx$$

with Z satisfying the residue condition. In fact

$$\frac{-\omega_{2i}Z}{W} \, dx$$

is holomorphic at s_i for $Z \in L_\Lambda^{para}(\mathcal{A})$. We can also show using the results in Part I, §4 that

$$\chi_{2i}(\gamma_i) = b_i \left(\int_{\overline{\gamma}_i} \frac{-\omega_{2i}Z}{W} \, dx \right)$$

where $\overline{\gamma}_i$ is the "straight" line path from x_0 to s_i gotten by contracting the small loop of γ_i around s_i :

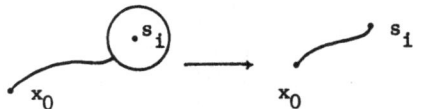

This is because near s_i the multivalued differential

$$\frac{\omega_{1i}Z}{W} \, dx$$

has logarithmic growth (see also Proposition I.4.8). Thus the normalized periods for γ_i are

$$[0, n_i] = [0, b_i(\int_{\overline{\gamma}_i} \frac{-\omega_{2i}Z}{W} \, dx + c_{1i})]$$

(note that n_i is uniquely determined up to sign independently of the choice of N_i),and the original periods for the basis ω_1, ω_2 and our fixed solution f are

$$[m_{\gamma_i}, n_{\gamma_i}] = [0, n_i]N_i \quad ,$$

which is easily seen to be

$$\left[b_i cd(\int_{\overline{\gamma_i}} \frac{-\omega_2 Z}{W} \, dx + c_1) - b_i c^2 (\int_{\overline{\gamma_i}} \frac{\omega_1 Z}{W} \, dx + c_2), \; b_i d^2 (\int_{\overline{\gamma_i}} \frac{-\omega_2 Z}{W} \, dx + c_1) \right.$$

$$\left. - b_i cd(\int_{\overline{\gamma_i}} \frac{\omega_1 Z}{W} + c_2) \right]$$

where $N_i = \begin{pmatrix} a & b \\ c & d \end{pmatrix}$. Note this is <u>independent</u> of the choice of N_i .

Notice that at each point $s_i \in S$ the periods $[m_{\gamma_i}, n_{\gamma_i}]$ depend on only one parameter, namely the normalized period n_i . The relation 3) introduces 2 distinct linear relations on the periods as is easy to see by choosing ω_1, ω_2 to give the normalized value around say γ_t and using the fact that the global monodromy group has finite index in $SL_2(\mathbb{Z})$. Thus our parabolic cocycle is determined by $\#S = t$ parameters subject to 2 relations. This gives a linear space of dimension $\#S-2$ which modulo the 2 dimensional space of coboundaries given by the choices of c_1, c_2 gives a space of dimension $\#S-4$ for the relevant parabolic cohomology group.

<u>Proposition V.1.1</u>: $\dim_{\mathbb{C}} L_\Lambda^{para}(\mathcal{O}\ell) \leq \#S-4$ and therefore $r =$

rank $E^{gen}(K(X)) \le \#S-4$ under our assumptions.   □

The choice of base point $x_0 \, \varepsilon \, X_0$ is certainly unimportant and it would be convenient instead to base our calculations at one of the points $s \, \varepsilon \, S$, say at s_1. Let's see how this can be done.

Fix a normalized K-basis ω_1, ω_2 at s_1 corresponding to E/X for the model 1) above. Thus the monodromy will be $\begin{pmatrix} 1 & b_1 \\ 0 & 1 \end{pmatrix}$ and as our local calculations in Part II show:

$$\omega_2(x) = a_r x^r + a_{r+1} x^{r+1} + \ldots \qquad a_r \ne 0$$

$$\omega_1(x) = \omega_2(x)\left[\frac{b_1}{2\pi i} \log x + \text{holomorphic}\right]$$

where x is assumed to be a good local parameter at s_1 and Λ has exponents (r,r) $r \, \varepsilon \, \mathbb{Z}$ at s_1. We remark that ω_2 is uniquely determined up to sign in this setting and ω_1 is unique up to the addition of an integer multiple of ω_2 and sign.

For $Z \, \varepsilon \, L_\Lambda^{para}(\mathcal{O})$ the inhomogeneous equation $\Lambda f = Z$ will have solutions of the form

$$f = \left(\int_{s_1}^x \frac{-\omega_2 Z}{W} \, dx + c_1\right)\omega_1 + \left(\int_{s_1}^x \frac{\omega_1 Z}{W} \, dx + c_2\right)\omega_2 \; .$$

The integrals make sense because for $Z \, \varepsilon \, L_\Lambda^{para}(\mathcal{O})$ the differential $\frac{-\omega_2 Z}{W} dx$ is holomorphic at s_1 and $\frac{\omega_1 Z}{W} dx$ has only logarithmic growth. In fact for $c_1 = 0$ the solutions will be single-valued meromorphic as is easy to see.

In this setting the periods at s_1 will be $[m_{\gamma_1}, n_{\gamma_1}] = [0, b_1 c_1]$ and for the other points $s_2, \ldots, s_t \, \varepsilon \, S$ will be given by

$$[m_{\gamma_i}, n_{\gamma_i}] = [0, n_i]N_i$$

as above with

$$n_i = b_i \left(\int_{\overline{\gamma}_i} \frac{-\omega_{2i} z}{W} + c_{1i} \right)$$

where $\overline{\gamma}_i$ is now a "straight line" path from s_1 to s_i (note this also works for $i = 1$ since the integral is zero and $c_{1i} = c_1$ in that case).

Thus to compute the periods we select paths $\overline{\gamma}_i$ from s_1 to s_i $i = 2, \ldots, t$ (which can be chosen to avoid the cosingular points in X_0):

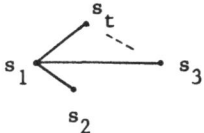

and compute the integrals above.

If we view these integrals as integrals on the universal cover of X_0', the paths would run from a fixed cusp (say at "$i\infty$") to inequivalent cusps (on the real axis) and could be taken as vertical lines provided we were able to avoid the cosingular points. We shall have more to say on this in a future paper (see also the remarks in §3 below). In the classical case where $X = \mathfrak{h}^*/\Gamma$ for $\Gamma \subset SL_2(\mathbb{Z})$ and E is the canonical elliptic modular surface over X, these periods are related to special values of the Dirichlet series attached to cusp forms of weight 3 (see Shimura [26]). In this sense the question of when a generalized automorphic form of weight 3 (which corresponds to a class in $H^2(E,\mathbb{C})$) represents an algebraic cycle is a rationality question about special values of an associated Dirichlet series.

Now the functional invariant J gives us a map $X \xrightarrow{J} \mathbb{P}^1_{\mathbb{C}}$ (note X is itself a sphere). The paths $\overline{\gamma}_i$ $i = 2, \ldots, t$ push down via J to paths $\overline{\tau}_i$ in $\mathbb{P}^1_{\mathbb{C}}$ from ∞ to ∞ which avoid 0, 1, the ramification points of $J: X \to \mathbb{P}^1_{\mathbb{C}}$, and places over which λ^2 has a pole or zero. The question we want to ask is whether or not we can express the integrals involved in computing the periods as "simple" integrals on $\mathbb{P}^1_{\mathbb{C}}$ over the paths $\overline{\tau}_i$. This is in fact the case and is the main result of this section. To do this we must examine a very special case.

Consider the u-sphere $\mathbb{P}^1_{\mathbb{C}}$ and the hypergeometric differential equation

4)
$$\frac{d^2 f}{du^2} + \frac{1}{u}\frac{df}{du} + \frac{31/144u - 1/36}{u^2(u-1)^2}\, f = 0 \quad .$$

The solution in terms of Riemann's P-function is

$$P \left\{ \begin{array}{ccc} 0 & \infty & 1 \\ -1/6 & 0 & 1/4 \\ 1/6 & 0 & 3/4 \end{array} \right\} = u^{-1/6}(u-1)^{1/4}\, P \left\{ \begin{array}{ccc} 0 & \infty & 1 \\ 0 & 1/12 & 0 \\ 1/3 & 1/12 & 1/2 \end{array} \right\}$$

which is seen to be a hypergeometric function. Thus at $u = 0$ we have two solutions

$$\eta_1 = u^{-1/6}(u-1)^{1/4}\; {}_2F_1\left(\tfrac{1}{12}, \tfrac{1}{12}; \tfrac{2}{3}; u\right)$$

$$\eta_2 = u^{1/6}\,(u-1)^{1/4}\; {}_2F_1\left(\tfrac{5}{12}, \tfrac{5}{12}; \tfrac{4}{3}; u\right)$$

which form a basis. We now let

$$c = (2 - \sqrt{3})\left[\frac{\Gamma(11/12)}{\Gamma(7/12)}\right]^2 \frac{\Gamma(2/3)}{\Gamma(4/3)}$$

(Γ the gamma function) and consider another basis of solutions at $u = 0$

$$\phi_1 = e^{2\pi i/3}\, \eta_1 + c\eta_2$$

$$\Phi_2 = \eta_1 - ce^{-\pi i/3}\eta_2 \quad .$$

The quotient of these solutions $\Phi(u) = \Phi_1(u)/\Phi_2(u)$ can be regarded as a multivalued function

$$\mathbb{P}_{\mathbb{C}}^1 - \{0,1,\infty\} \xrightarrow{\Phi} \mathbb{P}_{\mathbb{C}}^1 \quad .$$

The key point is that Φ is an inverse of the elliptic modular function J (Bateman [1]). Hence Φ maps $\mathbb{P}_{\mathbb{C}}^1 - \{0,1,\infty\}$ to $\hbar$ and Φ_1, Φ_2 form a K-basis of solutions for 4) which is then a K-equation.

To make the identification of $\Phi(u)$ with the inverse of the elliptic modular function more precise slit the unit disc $|u| < 1$ along the negative real axis

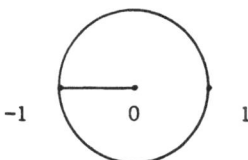

and use the principal branch of $\log u$ to define $u^{\pm 1/6}$, so that $u^{\pm 1/6}$ takes positive real values on $(0,1)$. We have on this slit disc that

$$\Phi(u) = \frac{e^{2\pi i/3} \, {}_2F_1\left(\frac{1}{12}, \frac{1}{12}; \frac{2}{3}; u\right) + cu^{1/3} \, {}_2F_1\left(\frac{5}{12}, \frac{5}{12}; \frac{4}{3}; u\right)}{{}_2F_1\left(\frac{1}{12}, \frac{1}{12}; \frac{2}{3}; u\right) - ce^{-\pi i/3} u^{1/3} \, {}_2F_1\left(\frac{5}{12}, \frac{5}{12}; \frac{4}{3}; u\right)} \quad .$$

Now $\lim_{u \to 0} \Phi(u) = e^{2\pi i/3}$ and because

$$\lim_{u \to 1} {}_2F_1\left(\frac{1}{12}, \frac{1}{12}; \frac{2}{3}; u\right) = \frac{\Gamma\left(\frac{2}{3}\right)\Gamma\left(\frac{1}{2}\right)}{\Gamma\left(\frac{7}{12}\right)\Gamma\left(\frac{7}{12}\right)}$$

$$\lim_{u \to 1} {}_2F_1\left(\frac{5}{12}, \frac{5}{12}; \frac{4}{3}; u\right) = \frac{\Gamma\left(\frac{4}{3}\right)\Gamma\left(\frac{1}{2}\right)}{\Gamma\left(\frac{11}{12}\right)\Gamma\left(\frac{11}{12}\right)}$$

an easy calculation gives

$$\lim_{u \to 1} \Phi(u) = i.$$

If we now slit the u-plane along the negative real axis and along the positive real axis from 1 to ∞:

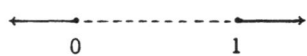

we can continue $\Phi(z)$ to a single-valued function which maps the slit plane onto a fundamental domain for $SL_2(\mathbb{Z})$ in $\mathfrak{h}$ as pictured below:

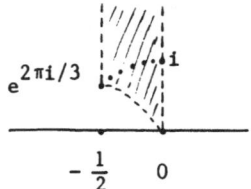

with the real axis between 0 and 1 mapping to the arc of the unit circle in $\mathfrak{h}$ running from $e^{2\pi i/3}$ to i.

For our purposes it will be necessary to understand the solutions of 4) at ∞. Let $s = \frac{1}{u}$ be a local parameter at ∞, then our differential equation in terms of s is

5)
$$\frac{d^2 f}{ds^2} + \frac{1}{s}\frac{df}{ds} + \frac{\left(\frac{31}{144} - \frac{1}{36}s\right)}{s(s-1)^2} f = 0$$

(see remarks following Theorem I.2.1). At $s = 0$ the exponents are 0,0 so our equation possesses a one dimensional space of single-valued holomorphic solutions near $s = 0$ (u = ∞). An easy calculation shows that these single-valued solutions are scalar multiples of

$$\psi_2(s) = \frac{\sqrt{2}}{3}\pi (1-s)^{1/4} \, _2F_1\left(\frac{1}{12}, \frac{5}{12}; 1; s\right)$$

(we take the (single-valued) branch of $(1-s)^{1/4}$ near $s = 0$ to be the one taking value 1 at 0). The factor $\frac{\sqrt{2}}{3}\pi$ is for later use. To find the other solution we suppress the equation and this leads to:

Proposition V.1.2: The solutions to 5) at $s = 0$ (i.e. to 4) at ∞) are of the form

$$\psi_2(s)\left[\int c_1\psi_2(s)^{-2}s^{-1}ds + c_2\right] \qquad c_1, c_2 \in \mathbb{C}$$

where $s = \frac{1}{u}$ is the local parameter at ∞. Note that $\psi_2(s)$ is holomorphic non-vanishing at $x = 0$ so invertible to a holomorphic non-vanishing function. $\qquad\qquad\qquad\qquad\qquad\qquad\qquad\qquad\qquad\qquad\qquad\square$

In particular if we choose $c_1 = -i\pi/9$, $c_2 = 0$ we get a solution

$$\psi_1(s) = \psi_2(s)\left(\frac{1}{2\pi i}\log s + \text{holomorphic vanishing}\right)$$

and $\psi_1(s)$, $\psi_2(s)$ form a basis of solutions at $s = 0$ ($u = \infty$) with monodromy $\begin{pmatrix} 1 & 1 \\ 0 & 1 \end{pmatrix}$. Now if we slit the unit s-disc, $|s| < 1$, along the negative real axis

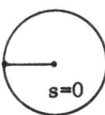

then because our equation gives the inverse to the J-function it must be possible to find two solutions to 5) at $s = 0$, call them $\omega_1(s)$, $\omega_2(s)$ whose quotient $\omega(s) = \omega_1(s)/\omega_2(s)$ maps the slit disc into a portion of the standard fundamental domain $\{\tau \in \mathcal{h}$ s.t. $-\frac{1}{2} < \text{Re } \tau < \frac{1}{2}$ and $|\tau| > 1\}$ near $\tau = "i\infty"$:

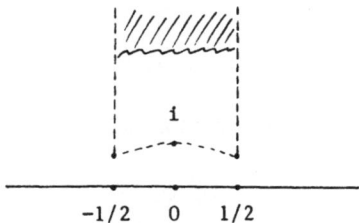

to provide a branch of the inverse to the J-function. Moreover the monodromy of ω_1, ω_2 will then necessarily be $\left(\begin{smallmatrix} 1 & 1 \\ 0 & 1 \end{smallmatrix}\right)$, so that $\omega_2(s)$ will be an invariant solution, i.e.

$$\omega_2(s) = \lambda\psi_2(s) \qquad \lambda \in \mathbb{C} - \{0\}$$

and ω_1 will be given as

$$\omega_1(s) = \lambda\psi_2(s)\left(\frac{1}{2\pi i} \log s + h + \text{holomorphic vanishing}\right)$$

$$= \lambda\psi_1(s) + h\lambda\psi_2(s)$$

for some $h \in \mathbb{C}$. Because we are interested at this point only in the quotient of solutions we can take $\lambda = 1$ so

$$\omega_2(s) = \psi_2(s)$$

$$\omega_1(s) = \psi_1(s) + h\psi_2(s)$$

and

$$\omega(s) = \frac{\psi_1(s)}{\psi_2(s)} + h$$

or more precisely

$$\omega(s) = -\frac{i\pi}{9} \int \omega_2(s)^{-2} s^{-1} ds + h$$

$$= \frac{1}{2\pi i} \log s + \left(-\frac{i\pi}{9}\right) \int_0^s \left(\omega_2(s)^{-2} - \frac{9}{2\pi^2}\right)s^{-1} ds + h$$

for some constant h. Note that the integrand $\left(\psi_2(s)^{-2} - \frac{9}{2\pi^2}\right)s^{-1}$ is holomorphic near $s = 0$. We need to determine h. To do this observe that we must have

$$\lim_{t \to 1} \omega(s) = i$$

because $J(i) = 1$ and $\omega(s)$ is supposed to be the inverse mapping into the standard fundamental domain for $SL_2(\mathbb{Z})$. Note that $\omega_2(s)$ is nonvanishing on the real axis between $s = 0$ and $s = 1$ and that

$$\lim_{t \to 1} {}_2F_1 \left(\tfrac{1}{12}, \tfrac{5}{12}; 1, s\right) = \frac{\Gamma(1)\Gamma(\tfrac{1}{2})}{\Gamma(\tfrac{11}{12})\Gamma(\tfrac{7}{12})} \neq 0 \quad .$$

Thus $\omega_2(s)^{-2} - \dfrac{9}{2\pi^2}$ grows like $(1-s)^{-1/2}$ as $s \to 1$. It follows the integral

$$-\frac{i\pi}{9} \int_0^1 \left(\omega_2(s)^{-2} - \frac{9}{2\pi^2}\right)s^{-1}\, ds$$

exists and that

$$h = i + \frac{i\pi}{9} \int_0^1 \left(\omega_2(s)^{-2} - \frac{9}{2\pi^2}\right)s^{-1}\, ds$$

which is a <u>purely imaginary number</u> because $\left(\omega_2(s)^{-2} - \dfrac{9}{2\pi^2}\right)s^{-1}$ is real valued on $(0,1)$. It is relatively straightforward to calculate that

$$h = \frac{3i}{2\pi} \log 12 \quad .$$

<u>Proposition V.1.3</u>: Let $\omega_2(s) = \psi_2(s)$ and $\omega_1(s) = \psi_1(s) + \dfrac{3i}{2\pi}(\log 12)\psi_2(s)$ as above. Then $\omega(s) = \omega_1(s)/\omega_2(s)$ is an inverse in a neighborhood of $u = \infty$ ($s = 0$) to the elliptic modular function $J(\tau)$ on $\mathcal{h}$. Moreover if we slit the u-plane from ∞ to 0 along the negative real axis and then from 0 to 1 along the real axis, $\omega(s)$ can be continued to a single-valued function which maps the slit sphere onto the standard fundamental domain for $SL_2(\mathbb{Z})$ in , i.e. $\{\tau \in \mathcal{h}, \ |\tau| > 1$ and $-\tfrac{1}{2} < \text{Re } \tau < \tfrac{1}{2}\}$. (Note this condition determines $\omega_1(s), \omega_2(s)$ up to a fixed scalar multiple.) □

We remark that the monodromy representation of our differential equation

$$\frac{d^2f}{du^2} + \frac{1}{u}\frac{df}{du} + \frac{\frac{31}{144}u - \frac{1}{36}}{u^2(u-1)^2} f = 0$$

with respect to the basis of solutions $\omega_1(s)$, $\omega_2(s)$ near $u = \infty$ ($s = 0$) is easily computed.

Slit the u-sphere from ∞ to 0 along the negative real axis and then slit from 0 to 1. The single-valued branch of the quotient of these solutions on this slit sphere take values in the usual fundamental domain $\{\tau \in \mathfrak{h}$ such that $-1/2 < \mathrm{Re}\ \tau < 1/2,\ |\tau| > 1\}$. Continuation across these slits in the directions indicated leads to the monodromy transformations shown (up to sign):

$$\pm\begin{pmatrix} 1 & 1 \\ 0 & 1 \end{pmatrix} \qquad 0 \qquad \pm\begin{pmatrix} 0 & 1 \\ -1 & 0 \end{pmatrix} \qquad 1$$

Taking γ_0, γ_1, γ_∞

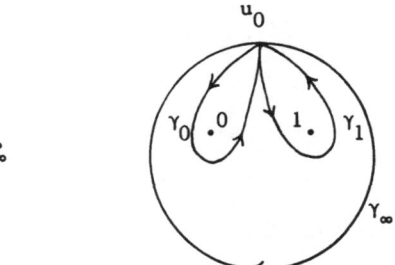

as generators for $\pi_1(\mathbb{P}^1_{\mathbb{C}} - \{0,1,\infty\}, u_0)$, we have:

$$\gamma_\infty \to \pm\begin{pmatrix} 1 & 1 \\ 0 & 1 \end{pmatrix}$$

$$\gamma_0 \to \pm\begin{pmatrix} 1 & 1 \\ -1 & 0 \end{pmatrix}$$

$$\gamma_1 \to \pm\begin{pmatrix} 0 & 1 \\ -1 & 0 \end{pmatrix} .$$

However at ∞ the exponents of 4) are 0,0 so the trace is +2 and at 0 the exponents are $-1/6$, $1/6$ so the trace is 1. Thus

$$\gamma_\infty \rightarrow \begin{pmatrix} 1 & 1 \\ 0 & 1 \end{pmatrix}$$

$$\gamma_0 \rightarrow \begin{pmatrix} 1 & 1 \\ -1 & 0 \end{pmatrix}$$

$$\gamma_1 \rightarrow \begin{pmatrix} 0 & -1 \\ 1 & 0 \end{pmatrix} .$$

This last because we must have

$$\begin{pmatrix} 1 & 1 \\ -1 & 0 \end{pmatrix}\begin{pmatrix} 0 & -1 \\ 1 & 0 \end{pmatrix}\begin{pmatrix} 1 & 1 \\ 0 & 1 \end{pmatrix} = \begin{pmatrix} 1 & 0 \\ 0 & 1 \end{pmatrix} .$$

We note for later use that the Wronskian $W = \omega_1(s)\dfrac{d}{ds}\,\omega_2(s) - \omega_2(s)\,\dfrac{d}{ds}\,\omega_1(s)$ $= \dfrac{i\pi}{9}\,s^{-1}$.

We now relate this differential equation to a particular elliptic surface. This example will play an important role in simplifying the period integrals above.

We consider the elliptic surface E over the u-sphere whose generic fiber is

8) $$Y^2 Z = 4X^3 - \frac{27u}{u-1}\,XZ^2 - \frac{27u}{u-1}\,Z^3 .$$

E is a basic elliptic surface and we take 8) as a model for the generic fiber E^{gen} over $\mathbb{C}(u)$. The functional invariant is $J = u$ (on the u-sphere). The differential equation which annihilates the periods of $\dfrac{dX}{Y}$ (see Stiller [29]) is:

$$\frac{d^2f}{du^2} + \frac{1}{u}\frac{df}{du} + \frac{(31/144)u - 1/36}{u^2(u-1)^2} f = 0$$

which is the hypergeometric equation we studied above. Further the singular fibers of $E/\mathbb{P}^1_{\mathbb{C}}$ are:

Type I_1 at ∞ $\quad\quad \begin{pmatrix} 1 & 1 \\ 0 & 1 \end{pmatrix}$

Type II at 0 $\quad\quad \begin{pmatrix} 1 & 1 \\ -1 & 0 \end{pmatrix}$

Type III* at 1 $\quad\quad \begin{pmatrix} 0 & -1 \\ 1 & 0 \end{pmatrix}$

(see Kodaira [16] for terminology). The matrix at the right is the local $SL_2(\mathbb{Z})$ matrix around such a singularity. This can be determined from the monodromy representation of the differential equation which we computed above.

Near $u = \infty$ we can arrange a basis $\tilde{\omega}_1(u)$, $\tilde{\omega}_2(u)$ for the periods of $\frac{dX}{Y}$ as functions of u (or $s = \frac{1}{u}$) so that the local monodromy around $u = \infty$ is $\begin{pmatrix} 1 & 1 \\ 0 & 1 \end{pmatrix}$ and Im $\tilde{\omega}_1/\tilde{\omega}_2 > 0$. Moreover because $J(\tilde{\omega}_1(u)/\tilde{\omega}_2(u))$ must equal the functional invariant u, $\tilde{\omega}_1(u)/\tilde{\omega}_2(u)$ will be an inverse for the elliptic modular function. Thus near $u = \infty$ ($s = 0$) the ratio $\tilde{\omega}_1(u)/\tilde{\omega}_2(u)$ is determined up to translation by $n \in \mathbb{Z}$ (J being invariant under such translations) and we are forced to have

$$\tilde{\omega}_2(u) = \lambda \omega_2(s)$$

$$\tilde{\omega}_1(u) = \lambda \omega_1(s) + \lambda n \omega_2(s)$$

for some $\lambda \in \mathbb{C} - \{0\}$, $n \in \mathbb{Z}$. But the periods of $\frac{dX}{Y}$ for a particular fiber are explicitly determined up to an $SL_2(\mathbb{Z})$ change of basis by our choice of model 8) for E^{gen} over $\mathbb{C}(u)$. Thus λ is actually determined (up to sign) and we shall seek to find its value.

At $u = \infty$ the fiber is

$$y^2 = 4x^3 - 27x - 27$$

which has an an ordinary double point at $\left(-\frac{3}{2}, 0\right)$. The group structure on the curve minus the double point is isomorphic to the multiplicative group $\mathbb{C}^*$ with $\frac{dX}{Y}$ being regular away from the double point. Thus $\frac{dX}{Y}$ has a single period at $u = \infty$ which corresponds to the "invariant" period at $u = \infty$. This period is thus

$$\lim_{u \to \infty} \tilde{\omega}_2(u) = \lim_{s \to 0} \lambda \omega_2(s) = \lambda \frac{\sqrt{2}}{3} \pi .$$

To evaluate this period we parameterize the rational curve $y^2 = 4x^3 - 27x - 27$ by

$$x = \frac{m^2}{4} + 3$$

$$y = \frac{m^3}{4} + \frac{9m}{2} = m\left(x + \frac{3}{2}\right)$$

which is one-to-one except that $m = \pm \sqrt{18} \, i$ map to the singularity $(-3/2, 0)$:

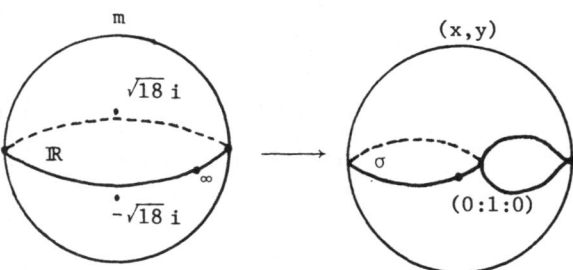

moreover the real points on the m-sphere including ∞ map to a loop σ (the loop of real points) as pictured, which generates the homology of the curve minus the double point. Thus up to sign the generating period of $\frac{dX}{Y}$ is

$$\int_\sigma \frac{dX}{Y} = \int_{-\infty}^\infty \frac{m/2}{\frac{m^3}{4} + \frac{9m}{2}} \, dm$$

$$= \int_{-\infty}^\infty \frac{1}{\left(\frac{m}{2}\right)^2 + 9} \, dm$$

$$= \frac{\sqrt{2}}{3} \tan^{-1}\left(\frac{m}{3\sqrt{2}}\right)\Big|_{-\infty}^\infty = \frac{\sqrt{2}\,\pi}{3}\,.$$

Note that $\dfrac{m/2}{m^3/4 + 9m/2}\, dm$ is holomorphic except at $\pm\sqrt{18}\,i$ where the residues are $\pm 1/\sqrt{18}\,i$. Thus $\lim\limits_{s\to 0} \lambda\omega_2(s) = \lambda\dfrac{\sqrt{2}}{3}\pi$ must be $\dfrac{\sqrt{2}}{3}\pi$ and therefore $\lambda = 1$. This explains our choice of constant $\dfrac{\sqrt{2}}{3}\pi$ earlier.

Proposition V.1.4: A basis for the periods of $\frac{dX}{Y}$ for the curve

$$Y^2 = 4X^3 - \frac{27u}{u-1} X - \frac{27u}{u-1} \quad \text{over} \quad \mathbb{C}(u) \quad \text{can be given by the functions}$$

$$\omega_1(s), \ \omega_2(s)$$

(as above) near $u = \infty$ ($s = 0$). Thus ω_1, ω_2 form a normalized basis for

the K-equation annihilating the periods of $\frac{dX}{Y}$ for 8). □

Let's return to our general setting. Let $E/X \cong \mathbb{P}^1_{\mathbb{C}}$ be a basic

elliptic surface having only multiplicative reduction as usual. We let

$\{s_1, \ldots, s_t\} = S \subset X$ denote the finite set of points where E/X has

singular fibers. We developed above some formulas for computing the

periods of the solutions to $\Lambda f = Z$ which are path integrals along paths

running from s_1 to the other points ("cusps") in S, i.e. $s_2, \ldots, s_t$.

In order to do these integrals we begin by selecting, near s_1, a

normalized basis ω_1, ω_2 for the periods of dX/Y of the chosen model

10) $$Y^2 = 4X^3 - g_2 X - g_3 \qquad g_2, g_3 \in K(X)$$

(normalized also means local monodromy $\begin{pmatrix} 1 & b_1 \\ 0 & 1 \end{pmatrix}$ at s_1). As the K-equation

$\Lambda_{(J,\lambda)}$ annihilates the periods of dX/Y (see Stiller [29]) where $\lambda^{-4} \frac{27J}{J-1}$

$= g_2$ and $\lambda^{-6} \frac{27J}{J-1} = g_3$, it is not hard to see that this normalized basis

ω_1, ω_2 on X is λ times the pull-back via J of the normalized basis

for 8) above, that is

$$\omega_i(x) = \lambda(\omega_i(u) \circ J)$$

where

$$X \overset{J}{\longmapsto} \mathbb{P}^1_{\mathbb{C}}$$

$$x \longmapsto u = J(x) \ .$$

Note that $\omega_1(x)$, $\omega_2(x)$ are unique up to the action of $\pm\begin{pmatrix} 1 & n \\ 1 & 0 \end{pmatrix}$ $n \in \mathbb{Z}$, so that ω_2 is unique up to sign and ω_1 up to an integer multiple of ω_2 and sign.

We now come to the final theorem of this section. Fix the model of $E^{gen}/K(X)$

$$Y^2 = 4X^3 - g_2 X - g_3$$

and at s_1 choose the normalized K-basis for this model

$$\omega_1 = \lambda(\omega_1 \circ J) \qquad \omega_2 = \lambda(\omega_2 \circ J)$$

(as above) where the K-equation annihilating the periods of dX/Y is $\Lambda = \Lambda_{(J,\lambda)}$. For $Z \in L_\Lambda^{para}(\mathcal{O}\!\ell) \subset L_\Lambda^{para}$ we consider the inhomogeneous equation $\Lambda f = Z$ near s_1.

In order to compute the periods we need to compute (see top page 109)

$$\int_{\overline{\gamma}_1} \frac{-(c\omega_1 + d\omega_2)Z}{W} \, dx$$

where $N_i = \begin{pmatrix} a & b \\ c & d \end{pmatrix}$ is as on page 107 and where $\overline{\gamma}_i$ is a path from s_1 to s_i and the continuation of $-\omega_{2i} = -(c\omega_1 + d\omega_2)$ along $\overline{\gamma}_i$ is an invariant solution at s_i.

<u>Theorem V.1.5:</u> $\displaystyle \int_{\overline{\gamma}_i} \frac{-(c\omega_1 + d\omega_2)Z}{W} \, dx$

$$= \frac{9i}{\pi} \int_{\tau_i} (c\omega_1(u) + d\omega_2(u))uZ' \, du \qquad Z' = \frac{\left(\frac{dx}{du}\right)^2 Z}{\lambda}$$

$$= -\frac{9i}{\pi} \int_{\overline{\tau}_i} (c\omega_1(s) + d\omega_2(s))sZ''ds \qquad Z'' = \frac{\left(\frac{dx}{ds}\right)^2 Z}{\lambda}$$

where ω_1, ω_2 are as above and $\overline{\tau}_i$ is the closed path on the u-sphere from ∞ to ∞ which is $J(\overline{\gamma}_i)$ where

$$J : X \longrightarrow \mathbb{P}^1_{\mathbb{C}}$$

$$x \longmapsto u = J(x),$$

and $s = 1/u$. We view Z' (or Z'') as an algebraic function of u (or s).

<u>Proof</u>: Consider a small neighborhood of s_1. Here

$$\frac{-(c\omega_1 + d\omega_2)Z}{W} dx = \frac{-c\lambda\omega_1(J(x)) - d\lambda\omega_2(J(x))Z}{\frac{i\pi}{9}\lambda^2 \frac{dJ/dx}{J}} \frac{dx}{dJ} dJ$$

$$= \frac{9i}{\pi}(c\omega_1(J(x)) + d\omega_2(J(x))) \frac{\left(\frac{dx}{dJ}\right)^2 Z}{\lambda} dJ.$$

If we now make the change of variables $u = J(x)$, which is valid along $\overline{\gamma}_i$ provided $\overline{\gamma}_i$ avoids points over the branch points of $X \xrightarrow{J} \mathbb{P}^1_{\mathbb{C}}$, then the result follows. We remark that the integrands in u or s have at worst logarithmic growth near $u = \infty$ or $s = 0$. $\qquad\square$

Note that the divisor $\mathcal{O}\!\ell$ in $L^{para}_\Lambda(\mathcal{O}\!\ell)$ depends on the derivation dx chosen to form the equation. If we change parameter to say dJ then the new divisor $\widetilde{\mathcal{O}\!\ell} = \mathcal{O}\!\ell + \mathrm{div}\left(\frac{dx}{dJ}\right)^2$ and we identify

$$L^{para}_\Lambda(\mathcal{O}\!\ell) \quad \text{with} \quad L^{para}_\Lambda(\widetilde{\mathcal{O}\!\ell})$$

via

$$Z \rightarrow Z\left(\frac{dx}{dJ}\right)^2 = \tilde{Z} \quad .$$

Thus we could simplify matters by starting with dJ as our derivation but this is often inconvenient.

Our problem finally comes down to computing a mixed hypergeometric-abelian integral around a closed path from ∞ to ∞ on the u-sphere. Before going on to some examples, we shall make a few remarks on these integrals from two other points of view which offer very interesting interpretations of these "periods". One involves the monodromy of normal functions and the other involves special values of Dirichlet series.

§2. Periods and the differential equations satisfied by normal functions.

In this section we shall give another useful interpretation of the periods. The setting will be identical to that of section §1 above; namely let E denote a basic elliptic surface having only multiplicative reduction over a curve X. We do not assume that E is regular, so $q = \dim_{\mathbb{C}} H^1(E, \mathcal{O}_E)$ need not be zero.

As always we fix a model

$$Y^2 = 4X^3 - g_2 X - g_3 \qquad g_2, \ g_3 \ \varepsilon \ K(X)$$

for the generic fiber of E/X. Let $\Lambda = \Lambda_{(J,\lambda)}$ be the corresponding K-equation as in §1. Taking $Z \ \varepsilon \ L_{\Lambda}^{para}(\mathcal{O})$, we consider the inhomogeneous equation

$$\Lambda_{(J,\lambda)} f = Z.$$

For a suitably chosen base point x_0 in $X_0 = X - S$, where S is the support of the singular fibers, and path $\gamma \ \varepsilon \ \pi_1(X_0, x_0)$ the periods $m_\gamma, \ n_\gamma$ are determined by the analytic (meromorphic) continuation of f around γ:

$$f \rightarrow f + m_\gamma \omega_1 + n_\gamma \omega_2.$$

Definition V.2.1: We shall refer to f above as a <u>normal function</u>. □

This terminology differs from the usual definition of a normal

function as a global section of a family of Jacobians (Zucker [38]).

However, when f has integral periods, and so comes from a section, the

two coincide. Recall also that the Leray spectral sequence for the

constant sheaf $\mathbb{C}$ on E over X degenerates at E_2 and that $L_\Lambda^{para}(\mathcal{O}l)$

can be identified with the first level F^1

$$F^1 \subset E_2^{1,1} = H^1(X, R^1\pi_*\mathbb{C})$$

of the Hodge filtration induced by the natural Hodge structure on $H^2(X,\mathbb{C})$;

see remarks in Part IV. We therefore have

$$F^1 = H^{2,0} \oplus H^{1,1} \quad \text{with} \quad H^{2,0} \cong H^0(E,\Omega_E^2),$$

and so our definition of normal function also permits us to look at the

periods associated to holomorphic two-forms, and to look at all the classes

in $H^{1,1}$ without restricting necessarily to rational or integral

classes.

Now if Z is any function in $L_\Lambda^{para}(\mathcal{O}l)$, the composite operator

$$\Lambda_Z = \left(\frac{d}{dx} - \frac{d}{dx} \log Z \right) \circ \Lambda$$

will be a third order linear homogeneous differential operator on X with

regular singular points whose only true singularities will be at the true

singularities of Λ. Observe that f, ω_1, ω_2 form a basis of solutions

for Λ_Z and that for $\gamma \in \pi_1(X_0,x_0)$ the monodromy takes the form:

$$\begin{pmatrix} f \\ \omega_1 \\ \omega_2 \end{pmatrix} \rightarrow \begin{pmatrix} 1 & m_\gamma & n_\gamma \\ 0 & a_\gamma & b_\gamma \\ 0 & c_\gamma & d_\gamma \end{pmatrix} \begin{pmatrix} f \\ \omega_1 \\ \omega_2 \end{pmatrix}$$

where m_γ, n_γ are the periods of f around γ as above and $\begin{pmatrix} a_\gamma & b_\gamma \\ c_\gamma & d_\gamma \end{pmatrix} \in$

$SL_2(\mathbb{Z})$ is the monodromy matrix for Λ with respect to ω_1, ω_2.

Definition V.2.2: Given a normal function f, so that

$$\Lambda f = Z, \quad Z \in L_\Lambda^{para}(\mathcal{H}),$$

we call Λ_Z above the <u>differential equation satisfied by the normal</u>

<u>function.</u>   □

Note that the monodromy representation

$$\gamma \to T_\gamma = \begin{pmatrix} 1 & m_\gamma & n_\gamma \\ 0 & a_\gamma & b_\gamma \\ 0 & c_\gamma & d_\gamma \end{pmatrix}$$

of the differential equation satisfied by the normal function f is (up to

equivalence) independent of the choice of solution to the equation

$\Lambda f = Z.$ For if we choose another particular solution $f + c_1 \omega_1 + c_2 \omega_2$ the

new monodromy matrix is

$$\begin{pmatrix} 1 & c_1 & c_2 \\ 0 & 1 & 0 \\ 0 & 0 & 1 \end{pmatrix} \begin{pmatrix} 1 & m_\gamma & n_\gamma \\ 0 & a_\gamma & b_\gamma \\ 0 & c_\gamma & d_\gamma \end{pmatrix} \begin{pmatrix} 1 & -c_1 & -c_2 \\ 0 & 1 & 0 \\ 0 & 0 & 1 \end{pmatrix}$$

$$= \begin{pmatrix} 1 & m_\gamma + c_1(a_\gamma-1) + c_2 c_\gamma & n_\gamma + c_1 b_\gamma + c_2(d_\gamma-1) \\ 0 & a_\gamma & b_\gamma \\ 0 & b_\gamma & d_\gamma \end{pmatrix}$$

and the periods transform to

$$[m_\gamma, \ n_\gamma] + [c_1, \ c_2](M_\gamma - I)$$

as they should (see Part I, §4 and Part II, §3).

In general not much is known about the monodromy representation of an

arbitrary 3rd (or nth) order linear homogeneous differential equation

with regular singular points on a Riemann surface X. This is an extremely

difficult but important problem. Fortunately there are a few cases where

something can be said, and as these will be useful to us in the examples,

we shall say a few words about them (see §4 below). In fact, our examples

also provide the solution to the monodromy problem in a number of new and
interesting cases.

Suppose for the moment that $\lambda = 1$ and that

$$X \xrightarrow{J} \mathbb{P}^1_{\mathbb{C}}$$
$$x \longrightarrow J(x) = u$$

is a cyclic cover ramified only over $0, 1, \infty$.

The function field $K(X)$ over $\mathbb{C}(u)$ $(u = J)$ is then a Galois

extension with cyclic Galois group. If Z is any function in $K(X)$ it

will satisfy a first order linear differential equation

1) $$(\frac{d}{du} - R_Z)Z = 0 \qquad R_Z = \frac{d}{du} \log Z \ \varepsilon \ \mathbb{C}(u)$$

on the u-sphere having regular singularities. In particular if

$Z \ \varepsilon \ L_\Lambda^{para}(\mathfrak{M})$, then because of our assumptions, the equation 1) for

$Z(\frac{dx}{du})^2 = Z(\frac{dx}{dJ})^2$ will have true singularities only at $0, 1, \infty$.

We wish to consider the third order equation

2) $$(\frac{d}{du} - R_{Z(\frac{dx}{dJ})^2}) \Lambda_{(u,1)}$$

where $\Lambda_{(u,1)}$ is the hypergeometric equation on the u-sphere discussed

in §1 above. The resulting third order equation has 3 true regular

singular points at $0, 1, \infty$. Our main result is that the periods of f

above are easily determined from the monodromy of this third order equation

on the u-sphere.

Theorem V.2.3: For a particular solution to the inhomogeneous equation

$\Lambda f = Z$ with $Z \ \varepsilon \ L_\Lambda^{para}(\mathfrak{M})$ the monodromy matrix of the associated third

order equation 2) for $Z(\frac{dx}{dJ})^2$ namely

$$(\frac{d}{du} - R_{Z(\frac{dx}{dJ})^2}) \Lambda_{(u,1)}$$

around the path $\tau = J \circ \gamma$ is

$$\begin{pmatrix} 1 & m_\gamma & n_\gamma \\ 0 & a_\gamma & b_\gamma \\ 0 & c_\gamma & d_\gamma \end{pmatrix}$$

where m_γ, n_γ are the periods of f and $\begin{pmatrix} a_\gamma & b_\gamma \\ c_\gamma & d_\gamma \end{pmatrix}$ is the monodromy

matrix for $\Lambda_{(J,1)}$ around γ.

<u>Proof</u>: Let $u_0 = J(x_0)$ be the base point on the u-sphere, and select a

K-basis ω_1, ω_2 at x_0 for $\Lambda_{(J,1)}$ corresponding to the periods of

dX/Y for the chosen model of E^{gen} over $K(X)$. Then as we have seen

$$\omega_1(x) = \omega_1(u) \circ J$$
$$\omega_2(x) = \omega_2(u) \circ J$$

where $\omega_1(u)$, $\omega_2(u)$ are a K-basis of solutions to $\Lambda_{(u,1)}$ at u_0

suitably normalized (see §1 above).

The function J is a good local parameter under our assumptions

except over 0, 1, ∞ on the u-sphere. If we express $\Lambda_{(J,1)}$ in terms

of the parameter J we get

$$\Lambda f = Z \left(\frac{dx}{dJ}\right)^2$$

(see page 7).

Thus our third order equation has for a basis of solutions at u_0 the

functions

$$\omega_1(u), \ \omega_2(u), \ f \ .$$

(Note $\Lambda_{(J,1)}$ is the pull-back of $\Lambda_{(u,1)}$.) Here f viewed on

the u-sphere satisfies

$$\Lambda_{(u,1)} f = Z \left(\frac{dx}{dJ}\right)^2$$

with $Z \left(\frac{dx}{dJ}\right)^2$ an algebraic function of u ($= J(x)$). We have taken here

the branch of Z corresponding to x_0. The result then follows as

$$\begin{pmatrix} f \\ \omega_1 \\ \omega_2 \end{pmatrix} \rightarrow \begin{pmatrix} 1 & m_\gamma & n_\gamma \\ 0 & a_\gamma & b_\gamma \\ 0 & c_\gamma & d_\gamma \end{pmatrix} \begin{pmatrix} f \\ \omega_1 \\ \omega_2 \end{pmatrix}.$$

Note $\begin{pmatrix} a_\gamma & b_\gamma \\ c_\gamma & d_\gamma \end{pmatrix} = \begin{pmatrix} a_\tau & b_\tau \\ c_\tau & d_\tau \end{pmatrix}$ and this block is the monodromy of the second

order hypergeometric equation $\Lambda_{(u,1)}$ around τ.   □

The point is that the equation

$$\left(\frac{d}{du} - R_{Z(\frac{dx}{dJ})^2} \right) \Lambda_{(u,1)}$$

has only regular singularities at three points and, just as with the

classical hypergeometric equation, its monodromy can sometimes be

determined in terms of the local parameters appearing in the differential

equation. This is particularly true when the equation is that of a

generlized hypergeometric function $_pF_q$ (see Bateman [1]).

If we relax our assumptions but still require $J: X \rightarrow \mathbb{P}^1_\mathbb{C}$ to be

ramified only over 0, 1, ∞; λ to have its zeros and poles only at the

points of X over 0, 1, ∞ in $\mathbb{P}^1_\mathbb{C}$; then for $Z \in L_\Lambda^{para}(\mathcal{O}t)$ we will have

$$\Lambda_{(J,1)}\left(\frac{f}{\lambda}\right) = \frac{Z}{\lambda}\left(\frac{dx}{dJ}\right)^2$$

where $\Lambda_{(J,1)}$ is expressed using J as a parameter. The algebraic

function

$$\frac{Z}{\lambda}\left(\frac{dx}{dJ}\right)^2$$

on the u-sphere will have to satisfy an nth-order differential equation

Π having true regular singularities only at 0, 1, ∞ and the differential

equation

$$\Pi \circ \Lambda_{(u,1)}$$

will be of order $n+2$ with regular singularities at 0, 1, ∞ and with f/λ, ω_1, ω_2 forming part of a basis of solutions on the u-sphere. Moreover the monodromy will be

$$
\pm \left(
\begin{array}{c|ccc}
 & & * & \\
\hline
0 \cdots 0 & 1 & m_\gamma & n_\gamma \\
0 \cdots 0 & 0 & a_\gamma & b_\gamma \\
0 \cdots 0 & 0 & c_\gamma & d_\gamma
\end{array}
\right)
$$

around suitable paths γ <u>on the u-sphere!</u>

This case arises in one very important class of examples. Namely let $\Gamma \in SL_2(\mathbb{Z})$ be a subgroup of finite index in $SL_2(\mathbb{Z})$ having no torsion and only regular cusps, then the canonical elliptic modular surface E_Γ over the modular curve, $X_\Gamma = \hat{h}^*/\overline{\Gamma}$ meets the requirements above for a suitable choice of λ. The $Z \in L_\Lambda^{para}(\mathcal{O}\!\!\iota)$ are then all of the first kind and correspond automorphic (cusp) forms of weight three. Thus there is an effective method for computing the periods of such a cusp form in terms of the monodromy of a differential equation and these periods are in turn related to special values of the associated Dirichlet series at certain integers (see Stiller [36]).

The general case (with no restrictions on the ramification) also gives the periods as entries in a monodromy matrix for a differential equation of higher order on the u-sphere. Unfortunately it may have more than three singularities making effective computation very difficult.

§3. <u>A formula, a method, and a remark on special values of Dirichlet</u>
<u>series.</u>

This brief section consists of three remarks that will assist in the understanding of the examples which follow.

The first remark concerns the map

$$0 \to E^{gen}(K(X))/torsion \to K(X)$$

$$\mathcal{P} \to \Lambda \int_{\mathcal{O}}^{\mathcal{P}} \frac{dX}{Y} = Z$$

which injects the group of $K(X)$-rational points on the generic fiber E^{gen}
into the additive group of $K(X)$. As usual we are assuming that the model
of E^{gen} is fixed

1) $Y^2 = 4X^3 - g_2 X - g_3$ $g_2, g_3 \in K(X)$

and that Λ annihilates the periods of $\frac{dX}{Y}$ where

$$\Lambda = \frac{d^2}{dx^2} + P\frac{d}{dx} + Q$$

for some "parameter" $x \in K(X)$. Now suppose we are given a solution
$(X(x), Y(x)) \in K(X)^2$ to 1) rational over $K(X)$, then we can compute Z
explicitly.

__Theorem V.3.1:__ Let $(X(x), Y(x))$ be a solution to

$$Y^2 = 4X^3 - g_2 X - g_3$$

rational over $K(X)$ and let $\Lambda_{(J,\lambda)}$ be the K-equation annihilating the
periods of $\frac{dX}{Y}$. Here $g_2 = \lambda^{-4}\frac{27J}{J-1}$ and $g_3 = \lambda^{-6}\frac{27J}{J-1}$. Then

$$Z = \Lambda_{(J,\lambda)} \int_{(0:1:0)}^{(X(x):Y(x):1)} \frac{dX}{Y}$$

$$= \frac{hX(x)^4 + \ell\lambda^{-4}X(x)^2 + m\lambda^{-6}X(x) + \lambda^{-8}n}{Y(x)^3}$$

$$+ P\frac{\frac{d}{dx}(\lambda^2 X(x))}{\lambda^2 Y(x)} + \frac{(\frac{d}{dx}(\lambda^2 X(x)))(s\lambda^2 X(x) + s)}{2\lambda^8 Y(x)^3}$$

$$+ \frac{\lambda^3 Y(x) \frac{d^2}{dx^2}(\lambda^2 X(x)) - \frac{d}{dx}(\lambda^3 Y(x)) \frac{d}{dx}(\lambda^2 X(x))}{\lambda^5 Y(x)^2}$$

where

$$P = \frac{\left(\frac{dJ}{dx}\right)^2 - J\left(\frac{d^2 J}{dx}\right)}{J \frac{dJ}{dx}}$$

$$s = \frac{-27 \frac{dJ}{dx}}{(J-1)^2}$$

$$h = -8\left(\frac{dJ}{dx}\right)^2 \left(\frac{31}{144} J - \frac{1}{36}\right) J^{-2}(J-1)^{-2}$$

$$\ell = \left(\frac{dJ}{dx}\right)^2 \left(\frac{87}{8} J - \frac{15}{2}\right) J^{-1}(J-1)^{-3}$$

$$m = \left(\frac{dJ}{dx}\right)^2 \left(\frac{51}{4} J - 6\right) J^{-1}(J-1)^{-3}$$

$$n = \frac{27}{8}\left(\frac{dJ}{dx}\right)^2 (J-1)^{-3}$$

and where λ is as above.

Proof: See Stiller [29].   □

Our second remark concerns the handling of singular fibers of type
other than I_b. These are the so-called additive types (additive
reduction) which divide into two classes, namely those with finite local
monodromy (I_0^*, III, III*, II, II*, IV, IV*) and those with infinite
cyclic local monodromy (I_b^*, b > 0). In either case, we can study the
local behaviour of our differential equations and normal functions by
passing to a finite cyclic cover (of order 2, 3, 4 or 6 appropriately)
where the fiber becomes either good or type I_{2b} respectively. This is
straightforward and permits us to extend all of our previous definitions
and results to the general case. Because we will be considering a number
of examples below which have some additive reduction, we thought it useful
to spend a little time here describing this type of behaviour.

As usual Λ will be a K-equation (or for that matter any second order equation with regular singular) points on X and $S \subset X$ will be the set of true singularities. We wish to consider inhomogeneous equations

$$\Lambda f = Z$$

with $Z \in L_\Lambda^{res}$ (see Definition II.3.2 -- note that L_Λ^{res} is defined without reference to the type of behaviour that occurs at the true singularities). Each choice of a particular solution f gives rise to a cocycle via analytic (meromorphic) continuation along paths in $X_0 = X-S$:

$$\gamma \in \pi_1(X_0, x_0) \longmapsto [m_\gamma, n_\gamma] \in \mathbb{C}^2$$

where f continues around γ as

$$f \longmapsto f + m_\gamma \omega_1 + n_\gamma \omega_2.$$

Two different choices of a particular solution give cocycles that differ by a coboundary. Thus each inhomogeneous equation

$$\Lambda f = Z \quad \text{with} \quad Z \in L_\Lambda^{res}$$

gives rise to a well-defined cohomology class, and it gives rise to the trivial class if and only if it is _exact_, i.e. there exists $Z' \in K(X)$ such that $\Lambda Z' = Z$, so that the equation has a _global_ single-valued solution.

The cohomology group in question is isomorphic to the usual group cohomology

$$H^1(\pi_1, V_{2,x_0})$$

where V_2 is the local system associated to Λ (in the case where Λ is a K-equation this is just the complexified homological invariant $G\big|_{X_0} \otimes_{\mathbf{Z}} \mathbf{C}$ as a sheaf of complex vector spaces on X_0) and where π_1 acts on the stalk V_{2,x_0} at the base point x_0 via monodromy in the usual way (see Deligne [5]). This cohomology group is also isomorphic to the cohomology of the local system as a sheaf of complex vector spaces

$$H^1(X_0, V_2)$$

which in turn is isomorphic to

$$\text{Ext}^1_{\underline{\mathbf{C}}}(V_1, V_2)$$

where V_1 is a trivial local system, so $V_1 \cong \underline{\mathbf{C}}$ the constant sheaf.

In this context, we can see how our inhomogeneous equations give rise to extensions

$$0 \to V_2 \to V_3 \to V_1 \to 0 .$$

V_3 is just the local system attached to the composite third order linear homogeneous equation

$$(\frac{d}{dx} - \frac{d}{dx} \log z) \circ \Lambda$$

with $0 \to V_2 \to V_3$ being the inclusion of the spaces of solutions and $V_3 \to V_1$ being the map induced by Λ. Moreover it is clear that every extension arises from such an inhomogeneous equation with $z \in L_\Lambda^{\text{res}}$ (see Stiller [36]).

We also will wish to consider inhomogeneous equations

$$\Lambda f = z \qquad z \in L_\Lambda^{\text{para}} \subset L_\Lambda^{\text{res}}$$

of the <u>second kind</u> for Λ a K-equation. Here L_Λ^{para} is defined as in

Definition II.3.3 by imposing an extra residue condition at those true

singular points $s \in S$ where the local monodromy is conjugate to $\left(\begin{smallmatrix} 1 & b \\ 0 & 1 \end{smallmatrix}\right)$.

Note that <u>no</u> conditions are imposed at the other types of singularities.

Second kind could also be called <u>locally exact</u> (more in the spirit of

Atiyah and Hodge [41]) since the definition is equivalent to the existence

of a single-valued solution in a neighborhod of each point of X. In terms

of extensions an inhomogeneous equation of the second kind

$$\Lambda f = Z \qquad Z \in L_\Lambda^{para}$$

gives rise to a locally split extension (see Stiller [36])

$$0 \to V_2 \to V_3 \to V_1 \to 0$$

which means that for every $x \in X$ there is a neighborhood U of x such

that on the punctured neighborhood the sequence

$$0 \to V_2\big|_{U-\{x\}} \to V_3\big|_{U-\{x\}} \to V_1\big|_{U-\{x\}} \to 0$$

splits, and conversely every locally split extension of V_1 by V_2 is

given by an inhomogeneous equation of the second kind. The cohomology

group

$$\frac{\text{second kind}}{\text{exact}} = \frac{\text{locally exact}}{\text{exact}} = \frac{\text{locally split extensions}}{\text{equivalence of extensions}}$$

is just the parabolic cohomology group H^1_{para} of Part II, §3 which is also

$H^1(X, R^1\pi_*\mathbb{C})$ where $\pi: E \to X$ is the projection.

Notice that any geometric normal function

$$f = \int_\theta^\varphi \frac{dX}{Y} \qquad \varphi \in E^{gen}(K(X))$$

is of the second kind as a consequence of Theorem III.2.7 (see remarks

following this theorem and Corollary III.2.8) and if $\wp$ is not torsion

$\Lambda \int_{\theta}^{\wp} \frac{dX}{Y} = \mathbb{Z}$ is not exact. In fact for a suitable multiple r of $\wp$ one

can show that

$$f = \int_{\theta}^{r\wp} \frac{dX}{Y} + m\omega_1 + n\omega_2$$

is single-valued locally for some choice of m, $n \in \mathbb{Z}$.

It remains to consider the filtration on

<u>second kind</u>
<u>exact</u>

(as in Part IV) for the cases with some additive reduction. This requires

proper definitions for the divisors $\mathcal{O}$ and $\mathcal{O}_0$ used in defining the

linear systems

$$L_\Lambda^{para}(\mathcal{O}) = L_\Lambda^{para} \cap L(\mathcal{O}) \subset L(\mathcal{O})$$

and

$$L(\mathcal{O}_0)$$

where the later is the space of things of the <u>first kind</u>.

In the case of $L(\mathcal{O}_0)$, first kind, we want Corollary III.5.9 to carry

over to the cases with additive reduction, that is, we want to be able to

identify $L(\mathcal{O}_0)$ with $H^0(E, \Omega_E^2)$ as before. This is straightforward (see

Table 1 below and the remarks which follow it) and is accomplished by

passing (locally) to an appropriately ramified cover. For example if the

local monodromy at $s \in S$ is finite of order e, we can pass to an e-fold

cover U of a disc V about s which is totally ramified at $\tilde{s}$ over

s. Pulling back our family $E|_{V-\{s\}}$ of elliptic curves gives a family

$\tilde{E}|_{U-\{\tilde{s}\}}$ on the cover with trivial local monodromy which can be compactified to a family $\tilde{E}\big|_U$ with good fiber at the point $\tilde{s}$ over s. $\tilde{E}\big|_U$ has on it a cyclic group of automorphisms of order e with quotient $E\big|_V$ except at s where the quotient has a single component with singularities whose minimal resolution gives $E\big|_V$. Any holomorphic two-form on $E\big|_V$ clearly pulls back to give a holomorphic two-form on $\tilde{E}\big|_U$ minus a finite number of points on the fiber over $\tilde{s}$ and thus extends to a holomorphic two-form on all of $\tilde{E}\big|_U$. Conversely, if

$$\Lambda f = Z \qquad Z \in L_\Lambda^{para}$$

pulls back to something of the first kind on the cover then it must have given rise to a holomorphic two-form on $E\big|_V$ to begin with. The actual construction of the two forms follows the previous method using the local holomorphic coordinates on the singular fibers given in Kodaira [16]. (The case of type IV* is covered in Shioda [27] explicitly.) Likewise one can handle the case of type I_b^*. In this way the divisor $\mathcal{O}_0$ is easily determined (see Theorem V.3.2 below).

The description of $\mathcal{O}$ and the linear systems $L_\Lambda^{para}(\mathcal{O}) \subset L(\mathcal{O})$ also proceeds by passing to the appropriately ramified cover. Here we use the observation that any section of $E\big|_V$ over V determines via "pull-back" a section of $\tilde{E}\big|_U$ over U.

In this way we can construct the desired Hodge filtration and extend all the results in Chapter IV to the additive case.

We can summarize everything in the tables below:

Table 1: Additive Fiber Types

r	s	type	k	e	①	②	ord W	i	remarks
1/6	$\frac{5}{6}+\ell$	II*	$\ell+\frac{2}{3}$	$3\ell+2$	≥-1	$\geq\ell-1$	ℓ	ℓ	
1/3	$\frac{2}{3}+\ell$	IV*	$\ell+\frac{1}{3}$	$3\ell+1$	≥-1	$\geq\ell-1$	ℓ	ℓ	
2/3	$\frac{4}{3}+\ell$	IV	$\ell+\frac{2}{3}$	$3\ell+2$	≥-1	$\geq\ell$	$\ell+1$	$\ell+1$	
5/6	$\frac{7}{6}+\ell$	II	$\ell+\frac{1}{3}$	$3\ell+1$	≥-1	$\geq\ell$	$\ell+1$	$\ell+1$	
1/4	$\frac{3}{4}+\ell$	III*	$\ell+\frac{1}{2}$	$2\ell+1$	≥-1	$\geq\ell-1$	ℓ	ℓ	
3/4	$\frac{5}{4}+\ell$	III	$\ell+\frac{1}{2}$	$2\ell+1$	≥-1	$\geq\ell$	$\ell+1$	$\ell+1$	
1/2	$\frac{3}{2}+\ell$	I_0^*	$\ell+1$	see rmks	≥-1	$\geq\ell$	$\ell+1$	$\ell+1$	$e=3\ell+3,2\ell+2,\ell+2,$ as $J=0,1$ or not
1/2	$\frac{1}{2}$	I_b^*	0	b	≥-1	≥-1	0	0	

Table 2: Multiplicative Fiber Types

0	0	I_b	0	b	≥-1	≥-1	-1	0	

Table 3: Good Fibers

0	1	good	1	see rmks	≥0	≥0	0	0	$e=3,2,1$ as $J=0,1$ or not
0	$2+\ell$	good but cosing.	$\ell+2$	see rmks	≥-1	$\geq\ell+1$	$\ell+1$	$\ell+1$*	$e=3\ell+6,2\ell+4,\ell+2$ as $J=0,1$ or not

*see Theorem V.3.3 below.

where

$\ell\in\mathbf{Z},\quad \ell\geq0$

$r\leq s$ are the exponents normalized so that $0\leq r<1$ by twisting Λ locally by a power of a local parameter t at the point $x\in X$ in question.

type is just the fiber type as designated by Kodaira in [16]

$k=s-r$ is the exponent difference

e is the ramification of J viewed as a map $X \xrightarrow{J} \mathbb{P}^1_{\mathbb{C}}$.

① indicates the order of pole for $Z \in L^{para}_\Lambda(\mathcal{O}l) \subset L(\mathcal{O}l)$ -- of course this is really the order of $t^{\bullet} Z (\frac{dx}{dt})^2$ because we have shifted to the local parameter t and twisted by a power of t, namely $t^{\bullet}$.

② indicates the order of pole (or zero) for $Z \in L(\mathcal{O}l_0)$ — again this is really the order of $t^{\bullet} Z (\frac{dx}{dt})^2$.

ord W is the order of the Wronskian

i is called the <u>index</u> — and as we shall see it represents the dimension contributed by the singular fiber to the (1,1)-part of

$$H^1(X, R^1 \pi_* \mathbb{C}) \quad \text{which is} \quad \tilde{=} \; L^{para}_\Lambda(\mathcal{O}l)/L(\mathcal{O}l_0).$$

For example, consider the first line of the table by supposing that Λ has lower exponent equal to $1/6 + n$ for some $n \in \mathbb{Z}$. We twist Λ by t^{-n} to get a new equation $\Lambda_{t^{-n}}$ with lower exponent $1/6$ at the point in question. It is trivial to see that the upper exponent of $\Lambda_{t^{-n}}$ will then be $5/6 + \ell$ for some non-negative integer ℓ and that this will only be for fiber type II*. Of course Λ itself will have upper exponent $5/6 + \ell + n$. The lines k and e are self-explanatory and ord W is the order of the Wronskian of $\Lambda_{t^{-n}}$ computed with respect to t. For column ① we want to consider what the order of Z will be for $Z \in L^{para}_\Lambda(\mathcal{O}l)$. Starting with

$$\Lambda f = Z \qquad Z \in L^{para}_\Lambda(\mathcal{O}l) \subset L(\mathcal{O}l)$$

we pass to a local parameter t at the point in question to get

$$\Lambda f = Z\left(\frac{dx}{dt}\right)^2$$

and then we twist (to normalize $r = 1/6$) getting

$$\Lambda_{t^{-n}} \, t^{-n} f = t^{-n} Z\left(\frac{dx}{dt}\right)^2.$$

The table indicates that

$$\mathrm{ord}\left(t^{-n} Z\left(\frac{dx}{dt}\right)^2\right) \geq -1 \ .$$

Thus at a point of type II*, $\mathcal{O}$ has order $1-n + \mathrm{ord}(dx)^2$ where n is the lower exponent of Λ minus $1/6$. To see how the entry ① is arrived at suppose that x is already a good local parameter and that Λ has exponents $1/6$, $5/6 + \ell$. Consider the branched cover

$$w \ \rightarrow w^6 = x$$

near the point in question. Pulling back and changing the parameter to w gives

$$\Lambda_w f(w) = Z(w)\left(\frac{dx}{dw}\right)^2$$

$$\Lambda_w f(w) = Z(w)(6w^5)^2.$$

Here the exponents of Λ_w are 1, $5+6\ell$ and we know that $Z(w)(6w^5)^2$ can have order at most 0 at a cosingular point with exponents 1, $5+6\ell$. It follows that $Z(w)$ has at most a 10th order pole and that therefore $Z(x)$ has order ≥ -1 $(-10/6)$. Column ② is similar. We find that at a point of type II*, $\mathcal{O}_0$ will have order $1 - \ell - n + \mathrm{ord}(dx)^2$ where $\ell+n$ is the upper exponent minus $5/6$.

Using this information we can prove a number of results.

__Theorem V.3.2:__ Consider the inhomogeneous equations of the first kind

$$\Lambda f = Z \qquad Z \in L(\mathcal{O\!l}_0)$$

then the degree of the divisor $\mathcal{O\!l}_0$ is easily computed to be

$$\deg \ \mathcal{O\!l}_0 = 2g-2 + \frac{\mu}{12} + \frac{5}{6} \, v(\text{II}\ast) + \frac{2}{3} \, v(\text{IV}\ast) + \frac{1}{3} \, v(\text{IV}) + \frac{1}{6} \, v(\text{II})$$

$$+ \frac{3}{4} \, v(\text{III}\ast) + \frac{1}{4} \, v(\text{III}) + \frac{1}{2} \, v(\text{I}_0^\ast) + \frac{1}{2} \sum_{b>0} v(\text{I}_b^\ast)$$

where g is the genus of X, μ is the valence of J, and $v(\ )$ the number of fibers of that type. Thus $\deg \mathcal{O\!l}_0 > 2g-2$ and as $p_g = \dim_{\mathbb{C}} H^0(E, \Omega_E^2) = \dim_{\mathbb{C}} L(\mathcal{O\!l}_0)$ it follows that

$$p_g = g - 1 + \frac{\mu}{12} + \frac{5}{6} \, v(\text{II}\ast) + \frac{2}{3} \, v(\text{IV}\ast) + \frac{1}{3} \, v(\text{IV}) + \frac{1}{6} \, v(\text{II})$$

$$+ \frac{3}{4} \, v(\text{III}\ast) + \frac{1}{4} \, v(\text{III}) + \frac{1}{2} \, v(\text{I}_0^\ast) + \frac{1}{2} \sum_{b>0} v(\text{I}_b^\ast).$$

This result was first proved by Kodaira in [16]. Our proof only makes use of the differential equation and thereby shows that generically isogeneous surfaces (which have the same differential equation) have the same geometric genus (see Stiller [30]).

__Proof:__ We shall begin by computing that the degree of the divisor $\mathcal{D}$ (Definition III.5.8) is $\dfrac{\text{valence } J}{12} - 2g + 2$, under the assumption of only multiplicative reduction. Let $\mu = \text{valence } J = $ degree of the map $X \xrightarrow{J} \mathbb{P}^1_{\mathbb{C}}$. We denote by $s_1, \ldots, s_r$ the points in S (which are the true singularities of Λ). At $s_i \in S$ the singular fiber will be of type I_{b_i} $b_i > 0$. Next let $p_1, \ldots, p_\ell$ be the points where $J = 0$, $q_1, \ldots, q_k$ the points where $J = 1$, and $r_1, \ldots, r_m$ the points where $J \neq 0, 1, \infty$ but $\text{ord } dJ > 0$. Say $\text{ord}_{r_i} dJ = n_i$. We denote by e_x the ramification index of $X \xrightarrow{J} \mathbb{P}^1_{\mathbb{C}}$ at $x \in X$.

The Hurwitz genus formula for $X \xrightarrow{J} P_{\mathbb{C}}^1$ gives:

$$2g - 2 + 2\mu = \sum_{x \in X}' (e_x - 1) = \sum_{i=1}^{r} (b_i - 1) + \sum_{i=1}^{m} n_i + (\mu - \ell) + (\mu - k) .$$

Thus

$$2g - 2 = \mu + \sum_{i=1}^{m} n_i - r - \ell - k.$$

On the other hand we can compute $-\sum_{x \in X}' (s_x - 1)$ directly by observing that this sum is independent of twisting Λ. We can thus assume Λ is of the form $\Lambda_{(J,1)}$ and use our local calculations from Part II, §2. We will have

$$\sum_{x \in X}' (-s_x + 1) = r - \sum_{i=1}^{m} n_i + \ell - \frac{\mu}{6} + k - \frac{3\mu}{4} .$$

Adding these two expressions gives

$$\sum_{x \in X}' (-s_x + 1) + 2g - 2 = \frac{\mu}{12} .$$

The computation of $\deg \mathcal{D}$ follows from this.

(This fact appears in Kodaira [16] with a considerably different proof -- as the irregularity q of the surface is equal to the genus g of X (see Part IV), the fact is equivalent to the statement that

$$\chi(\mathcal{O}_E) = p_g - q + 1 = \frac{\mu}{12} = \frac{\text{valence}}{12}$$

for surfaces having only singular fibers of type I_b.)

We shall follow the proof of this fact making the necessary changes to account for the case of additive reduction. As above let $\mu = \text{valence } J = $ degree of the map $X \xrightarrow{J} P_{\mathbb{C}}^1$, $r = $ the number of points where J has a pole, $\ell = $ the number of points where $J = 0$, $k = $ the number of points where $J = 1$, and denote by $r_1, \ldots, r_m$ the other ramification points where $J \neq 0, 1, \infty$ but $\text{ord } dJ > 0$. Say $\text{ord}_{r_i} dJ = n_i$. As before the Hurwitz formula gives

$$2g - 2 = \mu + \sum_{i=1}^{m} n_i - r - \ell - k \ .$$

We have also seen that

$$-\sum_{x \in X} (s_x - 1) = r - \sum_{i=1}^{m} n_i + \left(\ell - \frac{\mu}{6}\right) + \left(k - \frac{3\mu}{4}\right)$$

where s_x is the upper exponent of Λ at $x \in X$. Now

$$\mathfrak{A}_0 = \operatorname{div}(dx)^2 - \sum_{x \in X} t_x \cdot [x]$$

where $t_x \in \mathbf{Z}$ is determined by entry ② in the tables. For example at a fiber of type II* the differential equation would have exponents $1/6 + n$, $5/6 + \ell + n$ for some $n \in \mathbf{Z}$, and for Z of the first kind we would have

$$\operatorname{ord} Z\left(\frac{dx}{dt}\right)^2 \geq \ell + n - 1$$

where t is a local parameter at the point. Thus t_x is $\ell + n - 1$ which is $(5/6 + \ell + n) - 1 - 5/6 = (s_x - 1) - 5/6$. In each case

$$t_x = (s_x - 1) - \text{correction term}$$

with the correction term given by

$$\text{entry } s - \text{entry} \ ② - 1$$

where entry s, entry ② are the appropriate columns in the table. We can

summarize this

Type	Correction Term
II*	5/6
IV*	2/3
IV	1/3
II	1/6
III*	3/4
III	1/4
I_0^*	1/2
I_b^*	1/2
I_b	0
good	0

Thus it follows that

$$\deg \mathcal{O}_0 = 4g - 4 - \sum_{x \in X} (s_x - 1) + \frac{5}{6} v(II^*) + \frac{2}{3} v(IV^*)$$

$$+ \frac{1}{3} v(IV) + \frac{1}{6} v(II) + \frac{3}{4} v(III^*) + \frac{1}{4} v(III)$$

$$+ \frac{1}{2} v(I_0^*) + \frac{1}{2} \sum_{b>0} v(I_b^*) .$$

The result follows. □

We can also prove a result about the 1,1-part of the parabolic cohomology.

<u>Theorem V.3.3</u>: $\dim_{\mathbb{C}} H^1(X, R^1 \pi_* \mathbb{C}) - 2p_g = \text{rank } H^1(X,G) - 2p_g = \dim_{\mathbb{C}} L_\Lambda^{para}(\mathcal{O}) - p_g = \dim_{\mathbb{C}}(1,1)\text{-part of } H^1_{para} = \sum_{x \in X} i_x$ where i_x is the index given by the column headed i in the tables.

<u>Proof</u>: This follows from the Riemann-Roch theorem, the fact that $\mathcal{O} > \mathcal{O}_0$, and the fact that $\deg \mathcal{O}_0 > 2g - 2$. Thus $\dim_{\mathbb{C}} L(\mathcal{O}) - \dim_{\mathbb{C}} L(\mathcal{O}_0) = \dim_{\mathbb{C}} L(\mathcal{O}) - p_g = \sum_{x \in X} ② - ①$ where ② and ① are the appropriate table entries. In all cases this is i_x <u>except</u> at good fibers which are cosingular points where $② - ① = i_x + 1$. But for $L_\Lambda^{para}(\mathcal{O})$, each

cosingular point imposes one linear condition, namely the residue

condition. Note $Z \in L(\mathfrak{O})$ automatically satisfies the residue conditions

at the "cusps" of type I_b. □

Corollary V.3.4: (Shioda [27] and Cox and Zucker [4])

$$\text{rank } H^1(X,G) - 2p_g \geq v(I_0^*) + v(II) + v(III) + v(IV).$$

Proof: For each of these types $i_x \geq 1$. □

We can also characterize the elliptic modular surfaces of Shioda (see

Shioda [27]):

Theorem V.3.5: Let E/X be a basic elliptic surface and suppose that

$-\begin{pmatrix} 1 & 0 \\ 0 & 1 \end{pmatrix}$ is not in the global monodromy group of E/X and that $0 =$

rank $H^1(X,G) - 2p_g$, i.e. that

$$p_g = 2g - 2 + t - \frac{t_1}{2}$$

where t is the total number of singular fibers and t_1 the number of

type I_b $b > 0$, then X is a modular curve $M \backslash \mathfrak{h}^*$ for some subgroup

$M \subset SL_2(\mathbb{Z})$ of finite index with $-\begin{pmatrix} 1 & 0 \\ 0 & 1 \end{pmatrix} \notin M$ and E is the canonical

elliptic modular surface associated to M (Shioda [27]).

Proof: If $-\begin{pmatrix} 1 & 0 \\ 0 & 1 \end{pmatrix}$ is not in the global monodromy group then E can have

singular fibers of type I_b, I_b^* $b > 0$, IV and IV* only. Our other

hypothesis implies that $\sum_{x \in X} i_x = 0$ so that type IV does not occur and

there are no cosingular points. Let X_0 be X minus the "cusps", i.e.

the set of points where the fiber of E/X is singular of type I_b or I_b^*

$b > 0$. After fixing a suitable base point $x_0 \in X_0$ and picking a basis

ω_1, ω_2 of solutions for Λ, consider the resulting global monodromy group

$M \subset SL_2(\mathbb{Z})$ which is the image of the monodromy representation

$\pi_1(X_0 - \{\text{points of type IV*}\}, x_0) \to SL_2(\mathbb{Z})$ for ω_1, ω_2. We clearly have a map

(the period map):

$$X_0 \xrightarrow{\overline{\omega}} M\backslash \hbar$$

induced by $\omega = \omega_1/\omega_2$ which extends to

$$X \longrightarrow M\backslash \hbar^*$$

where $\hbar^* = \hbar \cup \mathbb{Q} \cup \{\infty\}$. By hypothesis $\overline{\omega}$ is unramified at $x \in X_0$
except over the elliptic points of the modular curve for $M\backslash \hbar$ where the
order of ramification will be one or three depending on whether or not the
fiber is type IV* or good respectively. Since $\hbar \to M\backslash \hbar$ ramifies to order
three over the elliptic points it is possible to lift this map to X_0:

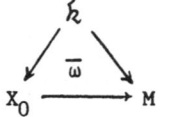

This immediately implies that there exists a subgroup M_0 of finite index
in M with $X_0 \cong M_0\backslash \hbar$ and that $\overline{\omega}$ is the canonical map

$$
\begin{array}{ccc}
\hbar & = & \hbar \\
\downarrow & & \downarrow \\
X_0 \cong M_0 & \xrightarrow{\overline{\omega}} & M\backslash \hbar .
\end{array}
$$

But as we have mentioned before, when $-I_2 \notin M$ the elliptic surface E is
induced by pulling back the elliptic modular surface for M via $\overline{\omega}$. Thus
our diagram would imply that E was the elliptic modular surface for M_0
which has monodromy M_0 -- therefore $M_0 = M$ and $X \cong M\backslash \hbar^*$ with E
being the elliptic modular surface for M. (Note if type IV* doesn't occur
either, then the proof is very easy, because the period map ω is
unramified

$$
\begin{array}{ccc}
\mathcal{h} & \xrightarrow{\ \omega\ } & \mathcal{h} \\
\downarrow & & \downarrow \\
X_0 & \longrightarrow & M
\end{array}
$$

and therefore an automorphism of $\mathcal{h}$.)   □

Our final remark concerns an interpretation of the periods as special values of certain Dirichlet series. For simplicity we shall restrict ourselves to the case where E has only multiplicative reduction. As always we denote the support of the singular fibers by $S = \{s_1,\ldots,s_t\} \subset X$ and its complement $X-S$ by X_0. In section one above we showed how to express the periods in terms of the integrals

$$
\int_{\overline{\gamma}_i} \frac{-\omega_2 Z}{W}\, dx \quad\text{and}\quad \int_{\overline{\gamma}_i} \frac{\omega_1 Z}{W}\, dx \qquad Z \in L_\Lambda^{para}(\mathcal{O}l)
$$

for $i = 2,\ldots,t$ with $\overline{\gamma}_i$ being a path from s_1 to s_i which can be taken to avoid the cosingular points in X_0.

In particular, fix a base cusp s_1 and let s_i be one of the other cusps. We can then choose $\Gamma \subset PSL_2(\mathbb{R})$ uniformizing X_0 in such a way that $i\infty$ is a cusp of Γ over s_1 and 0 is a cusp of Γ over s_i:

$$
\begin{array}{c}
\mathcal{h} \\
\downarrow \\
X_0 \overset{\sim}{=} \Gamma\backslash \mathcal{h}.
\end{array}
$$

Now let $\overline{\gamma}_i$ correspond to the image of the imaginary axis (with allowances made for going around the cosingular points) under our uniformization. We then have

$$
\int_{\overline{\gamma}_i} \frac{-\omega_2 Z}{W}\, dx = \int_{\overline{\gamma}_i} \frac{\omega_2^3 Z}{W^2} \cdot \frac{-W}{\omega_2^2}\, dx = \int_0^{1\infty} -g\, d\omega
$$

and

$$\int_{\overline{\gamma}_1} \frac{\omega_1 Z}{W}\, dx = \int_0^{i\infty} g\omega d\omega$$

where g is the associated generalized automorphic form. In addition we have:

<u>Proposition V.3.6</u>: Let $\Lambda f = Z$, $Z \in L_\Lambda^{para}(\mathcal{O}l)$, be an inhomogeneous equation with associated generalized automorphic form g, then g has a <u>canonical</u> series expansion at i^∞ as:

$$g(z) = \sum_{n \geq 1}^{\infty} a_n\, e^{2\pi i n \omega(z)/b}$$

where the fiber type at s_1 is I_b and $\omega = \omega_1/\omega_2$ is given by a normalized basis ω_1, ω_2 at s_1.

<u>Proof</u>: Using Theorem III.3.1 it is enough to see that $e^{2\pi i \omega(z)/b}$ is a local parameter at $s_1 \in X$, but this is obvious from the local form of the normalized solutions of Λ at s_1:

$$\omega_1 = \omega_2\left(\frac{b}{2\pi i}\, \log t + \text{holomorphic}\right)$$

where t is a local parameter at s_1. $\square$

   This result is equivalent to the existence of a fractional power series expansion in terms of the usual local parameter $q = e^{2\pi i \tau/\ell}$ at i^∞ for the global monodromy group $M \subset SL_2(\mathbb{Z})$ for some $\ell | b$.

   Thus formally at least we have a Dirichlet series

$$L_g(s) = \sum_{n \geq 1}^{\infty} \frac{a_n}{n^s}$$

and the integrals above come from a "Mellin-transform"

$$i^s M_g(s) = \int_0^\infty g\omega^{s-1} d\omega \ .$$

In this sense they represent (up to the usual exponential and gamma factors) the special values

$$L_g(1) \quad \text{and} \quad L_g(2).$$

Unfortunately the relationship is at present only conjectural except when g is of the first kind and sufficiently holomorphic on $\hbar$ (see Stiller [36]). One can also use our descriptions to calculate the special values for some classical cusp forms. The book of Exton [40] on hypergeometric integrals is particularly useful for such calculations.

Finally we remark that the integrals can be realized as actual Mellin-transforms

$$\int_0^\infty g(it)t^{s-1} \, dt$$

on $\hbar$ over the modular curve $\overline{M}\backslash\hbar$, except one must allow for the possibility of conjugating so that $M \subset GL_2^+(\mathbb{Q})$ commensurate with $SL_2(\mathbb{Z})$.

§4. Examples.

In this section we shall make use of the results in the previous sections to compute the Picard numbers of certain elliptic surfaces. Example 1: Our first two examples are extremely simple, but serve to illustrate the method and to establish the format that we shall follow in presenting the other examples.

Consider over the x-sphere $\mathbb{P}^1_{\mathbb{C}}$ the surface E given by

1) $$E^{gen}: Y^2 = 4X^3 - (3x^2(x^2-1)^3)X - (x^2(x^2-1)^5)$$

which has functional invariant

$J = x^2$ and $\lambda^2 = \dfrac{3}{(x^2-1)^2}$. The K-equation annihilating the periods of $\dfrac{dX}{Y}$

for this model of E^{gen} is

$$\Lambda = \frac{d^2}{dx^2} + \frac{5x^2-1}{x(x^2-1)}\frac{d}{dx} + \frac{4x^4 - 3\frac{5}{36}x^2 - \frac{1}{9}}{x^2(x^2-1)^2}$$

which yields the following data: (see the tables on page 140)

point x =	exponents	fiber type	ord Z, $Z\varepsilon L_\Lambda^{para}(\mathcal{O}\!\ell)$	ord Z, $Z\varepsilon L(\mathcal{O}\!\ell_0)$
0	$-\frac{1}{3}$, $+\frac{1}{3}$	IV	≥ -2	≥ -1
-1	$-\frac{3}{4}$, $-\frac{1}{4}$	III*	≥ -2	≥ -2
1	$-\frac{3}{4}$, $-\frac{1}{4}$	III*	≥ -2	≥ -2
∞	2 , 2	I_2	≥ 5	≥ 5
elsewhere	0 1	good	≥ 0	≥ 0

from which we see that any normal function will satisfy an inhomogeneous

equation $\Lambda f = Z$ with Z in

$$L_\Lambda^{para}(\mathcal{O}\!\ell) = L(\mathcal{O}\!\ell) = L(+2[0] + 2[1] + 2[-1] - 5[\infty]) = \{\frac{Ax+B}{x^2(x^2-1)^2} \; ; \; A,B \in \mathbb{C}\},$$

and that the elements of the first kind are given by Z in

$$L(\mathcal{O}\!\ell_0) = L(+1[0] + 2[1] + 2[-1] - 5[\infty]) = \{\frac{Ax}{x^2(x^2-1)^2} \; ; \; A \in \mathbb{C}\} \; .$$

We also see that E is an elliptic $K3$ surface with Picard number $\rho = 19$

or 20. We shall show below in Prop. V.4.2 that $\rho = 20$.

Before considering this, we shall further analyze our surface by

determining its monodromy representation. Let $t = \dfrac{1}{x}$ be a local parameter

at $x = \infty$ and consider the t-plane minus the imaginary axis and the two

rays $|t| \geq 1$ on the real axis

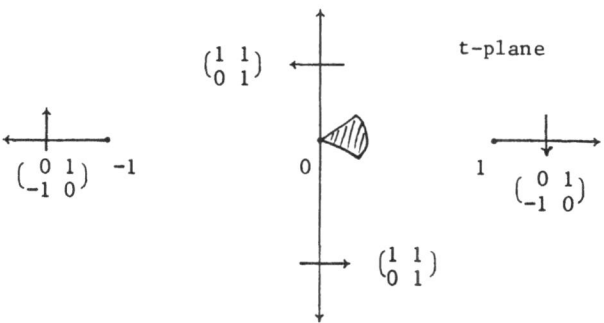

The pull-back via $t \to t^2 = s$ of the standard solutions $\omega_1(s)$, $\omega_2(s)$ twisted by λ (see Proposition V.1.3) gives a basis for the solutions of Λ on the slit (disconnected) t-plane which continues across the slits as shown above. Choosing this basis in a sector at 0 on the positive real side of the imaginary axis and a basis for $\pi_1(\mathbb{P}^1_{\mathbb{C}} - \{0, \pm 1, \infty\})$ as shown:

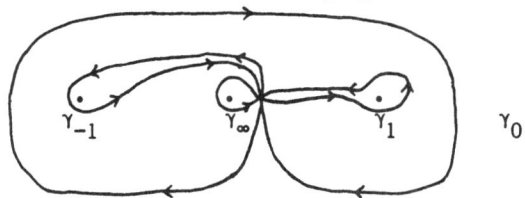

where γ_∞ goes around $x = \infty$ (i.e. $t = 0$) once in a counterclockwise manner etc., we get the monodromy representation:

$$\gamma_\infty \longmapsto \begin{pmatrix} 1 & 2 \\ 0 & 1 \end{pmatrix} = M_{\gamma_\infty}$$

$$\gamma_0 \longmapsto \begin{pmatrix} 0 & 1 \\ -1 & -1 \end{pmatrix} = M_{\gamma_0}$$

$$\gamma_1 \longmapsto \begin{pmatrix} 0 & -1 \\ 1 & 0 \end{pmatrix} = M_{\gamma_1}$$

$$\gamma_{-1} \longmapsto \begin{pmatrix} 1 & -2 \\ 1 & -1 \end{pmatrix} = \begin{pmatrix} 1 & 1 \\ 0 & 1 \end{pmatrix}\begin{pmatrix} 0 & -1 \\ 1 & 0 \end{pmatrix}\begin{pmatrix} 1 & -1 \\ 0 & 1 \end{pmatrix} = M_{\gamma_{-1}}.$$

<u>Proposition V.4.1</u>: The global monodromy group of E is $M = SL_2(\mathbb{Z})$ and the Mordell-Weil group $E^{gen}(K(X)) = E^{gen}(\mathbb{C}(x))$ is torsion-free. (This is also clear from the fiber types.) □

We now consider the specific basis of solutions $\omega_1(t)$, $\omega_2(t)$ of Λ
(refer to pp. 117-120) defined in the slit disc $|t| < 1$ minus the negative
imaginary axis

$$\omega_2(t) = \frac{\sqrt{2}\,\pi}{\sqrt{3}}\, t^2(1 - t^2)^{-3/4} \, {}_2F_1\left(\frac{1}{12}, \frac{5}{12}; 1; t^2\right)$$

$$\omega_1(t) = \omega_2(t)\left(\frac{1}{\pi i} \log t + \frac{31}{2\pi} \log 12 + \text{holomorphic vanishing}\right)$$

where $-\pi/2 < \arg t < 3\pi/2$ in our branch of the log and the root
$(1-t^2)^{-3/4}$ is taken as positive real on the positive real axis. This
basis is a normalized basis in the sense of section 1 above, giving
precisely the periods of $\frac{dX}{Y}$. It clearly analytically continues to the
whole t-plane minus the negative imaginary axis and the two rays $|t| \geq 1$
on the real axis with monodromy across the slits as shown:

t

1 bis)

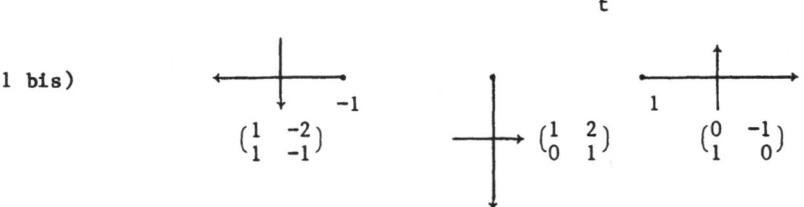

$$\begin{pmatrix} 1 & -2 \\ 1 & -1 \end{pmatrix} \qquad \begin{pmatrix} 1 & 2 \\ 0 & 1 \end{pmatrix} \qquad \begin{pmatrix} 0 & -1 \\ 1 & 0 \end{pmatrix}$$

Now let $Z \in L_\Lambda^{\text{para}}(\mathcal{O}t)$ so $Z = \dfrac{Ax+B}{x^2(x^2-1)^2}$ $A, B \in \mathbb{C}$. We wish to consider
the inhomogeneous equation

$$\Lambda f = Z$$

and in particular, we wish to consider it near $x = \infty$, i.e. $t = 0$. In
order to do this we change parameter which yields

$$\Lambda_t = \frac{d^2}{dt^2} + \frac{-3-t^2}{t(1-t^2)} \frac{d}{dt} + \frac{4 - 3\frac{5}{36} t^2 - \frac{1}{9} t^4}{t^2(1 - t^2)^2}$$

$$Z_t = Z\left(\frac{dx}{dt}\right)^2 = \frac{At + Bt^2}{(1-t^2)^2}$$

and f will satisfy the equation $\Lambda_t f = Z_t$, which is

$$\frac{d^2f}{dt^2} + \frac{-3-t^2}{t(1-t^2)} \frac{df}{dt} + \frac{4 - 3\frac{5}{36}t^2 - \frac{1}{9}t^4}{t^2(1-t^2)^2} f = \frac{At + Bt^2}{(1-t^2)^2} \quad .$$

As we have seen many times

$$f = \left(\int_0^t \frac{-\omega_2 z_t}{W} \, dt + c_1\right)\omega_1(t) + \left(\int_0^t \frac{\omega_1 z_t}{W} \, dt + c_2\right)\omega_2(t)$$

which is single-valued on the slit t-plane described in 1 bis) above. We
are interested in the periods of the normal function f. This is, as
explained in section 2, essentially the monodromy of the third order
equation

$$\left(\frac{d}{dt} - \frac{d}{dt} \log z_t\right) \circ \Lambda_t$$

which has f, ω_1, ω_2 as a basis of solutions. Its monodromy takes the
form

$$\gamma_\infty \to T_{\gamma_\infty} = \begin{pmatrix} 1 & m_{\gamma_\infty} & n_{\gamma_\infty} \\ 0 & & \\ 0 & & M_{\gamma_\infty} \end{pmatrix} = \begin{pmatrix} 1 & 0 & 2c_1 \\ 0 & 1 & 2 \\ 0 & 0 & 1 \end{pmatrix}$$

$$\gamma_1 \to T_{\gamma_1} = \begin{pmatrix} 1 & m_{\gamma_1} & n_{\gamma_1} \\ 0 & & \\ 0 & & M_{\gamma_1} \end{pmatrix} = \begin{pmatrix} 1 & m_{\gamma_1} & n_{\gamma_1} \\ 0 & 0 & -1 \\ 0 & 1 & 0 \end{pmatrix}$$

$$\gamma_{-1} \to T_{\gamma_{-1}} = \begin{pmatrix} 1 & m_{\gamma_{-1}} & n_{\gamma_{-1}} \\ 0 & 1 & -2 \\ 0 & 1 & -1 \end{pmatrix}$$

$$\gamma_0 \to T_{\gamma_0} = \begin{pmatrix} 1 & m_{\gamma_0} & n_{\gamma_0} \\ 0 & 0 & 1 \\ 0 & -1 & -1 \end{pmatrix} = T_{\gamma_\infty}^{-1} T_{\gamma_{-1}}^{-1} T_{\gamma_1}^{-1} =$$

$$= \begin{pmatrix} 1 & n_{\gamma_1}+2m_{\gamma_{-1}}+n_{\gamma_{-1}}+2c_1 & -m_{\gamma_1}+m_{\gamma_{-1}}+n_{\gamma_{-1}}+2c_1 \\ 0 & 0 & 1 \\ 0 & -1 & -1 \end{pmatrix}$$

so $m_{\gamma_0} = n_{\gamma_1} + 2m_{\gamma_{-1}} + n_{\gamma_{-1}} + 2c_1$ and $n_{\gamma_0} = -m_{\gamma_1} + m_{\gamma_{-1}} + n_{\gamma_{-1}} + 2c_1$.

We want to know for which $Z \in L_\Lambda^{para}(\mathcal{O})$ the inhomogeneous equation $\Lambda f = Z$ will have periods in $\mathbb{Z}$; for then the function f will be of the form

$$f = \int_{\mathcal{O}}^{\mathcal{P}} \frac{dX}{Y}$$

(up to integral multiples of the periods) for some $\mathcal{P} \in E^{gen}(\mathbb{C}(x))$ corresponding to a section of $E/X = \mathbb{P}_{\mathbb{C}}^1$. We have:

__Proposition V.4.2:__ $\Lambda f = Z$ for $Z \in L_\Lambda^{para}(\mathcal{O}) = \{\dfrac{Ax + B}{x^2(x^2-1)^2} \; A, B \in \mathbb{C}\}$ has integral periods if and only if $A = 0$ and $B = -\dfrac{in}{3\sqrt{3}}$ for $n \in \mathbb{Z}$. It follows that $E^{gen}(\mathbb{C}(x)) \cong \mathbb{Z}$ and that E has Picard number 20.

__Proof:__ Since the rank of the Mordell–Weil group is at most one, it suffices to show that

$$\Lambda f = \frac{B}{x^2(x^2-1)^2}$$

has integral periods if and only if $B = \dfrac{-in}{3\sqrt{3}}$, $n \in \mathbb{Z}$. So given such an equation, we want to know when we can choose c_1, c_2 (mod $\mathbb{Z}$) so that the particular solution determined by this choice has integral periods. Thus we want to choose c_1, c_2 so that $2c_1$, m_{γ_1}, n_{γ_1}, $m_{\gamma_{-1}}$ and $n_{\gamma_{-1}}$ are in $\mathbb{Z}$. Now from our explicit description of the periods we see that (refer to p. 112)

$$m_{\gamma_1} = -\int_0^1 \frac{-\omega_2 Z_t}{W} \, dt - c_1 + \int_0^1 \frac{\omega_1 Z_t}{W} \, dt + c_2$$

$$n_{\gamma_1} = -\int_0^1 \frac{-\omega_2 Z_t}{W} \, dt - c_1 - \int_0^1 \frac{\omega_1 Z_t}{W} \, dt - c_2$$

$$m_{\gamma_{-1}} = \int_0^{-1} \frac{\omega_1 Z_t}{W} \, dt + c_2$$

$$n_{\gamma_{-1}} = -2 \int_0^{-1} \frac{-\omega_2 Z_t}{W} \, dt - 2c_1 - 2 \int_0^{-1} \frac{\omega_1 Z_t}{W} \, dt - 2c_2$$

where the paths of integration are taken along the real axis. Note that

$$W = \omega_1(t) \frac{d}{dt} \omega_2(t) - \omega_2(t) \frac{d}{dt} \omega_1(t) = \frac{2\pi i}{3} \cdot \frac{t^3}{(1-t^2)^2}$$

so that

$$\frac{-\omega_2(t) Z_t}{W} \, dt = \frac{\sqrt{3} \, iB}{\sqrt{2}} t(1-t^2)^{-3/4} \, {}_2F_1\left(\frac{1}{12}, \frac{5}{12}; 1; t^2\right) dt$$

$$\frac{\omega_1(t) Z_t}{W} dt = \frac{-\sqrt{3} \, iB}{\sqrt{2}} t(1-t^2)^{-3/4} \, {}_2F_1\left(\frac{1}{12}, \frac{5}{12}; 1, t^2\right)\left(\frac{1}{\pi i}\log t + \text{holomorphic in } t^2\right) dt$$

and we see that all the integrals above converge. Finally under $t \to -t$, we see that

$$\int_0^{-1} \frac{\omega_1 Z_t}{W} \, dt = \int_0^{1} \frac{\omega_1 Z_t}{W} \, dt - \int_0^{1} \frac{-\omega_2 Z_t}{W} \, dt$$

$$\int_0^{-1} \frac{-\omega_2 Z_t}{W} \, dt = \int_0^{1} \frac{-\omega_2 Z_t}{W} \, dt$$

so

$$m_{\gamma_{-1}} = \int_0^{1} \frac{\omega_1 Z_t}{W} \, dt + c_2 - \int_0^{1} \frac{-\omega_2 Z_t}{W} \, dt$$

$$n_{\gamma_{-1}} = -2 \int_0^{1} \frac{\omega_1 Z_t}{W} \, dt - 2c_2 - 2c_1 \, .$$

Now we want to choose c_1, c_2 (mod $\mathbf{Z}$) so that first of all $2c_1 \in \mathbf{Z}$ — so without loss of generality we can assume $c_1 = 0$ or $1/2$. In the case $c_1 = 0$ the periods can be made integral if and only if

$$X_1 = \int_0^1 \frac{-\omega_2 Z_t}{W} \, dt \ \varepsilon \ \frac{1}{2} \, \mathbb{Z}$$

by choosing $c_2 = -\int_0^1 \frac{\omega_1 Z_t}{W} \, dt + \frac{1}{2} \bmod \mathbb{Z}$ if $X_1 \notin \mathbb{Z}$ or $c_2 = -\int_0^1 \frac{\omega_1 Z_t}{W} \, dt$ $\bmod \mathbb{Z}$ if $X_1 \varepsilon \mathbb{Z}$. In the case $c_1 = 1/2$ the periods can't be made integral as $m_{\gamma_1} - m_{\gamma_{-1}} = -c_1$.

Thus the periods can be made integral if and only if

$$\int_0^1 \frac{-\omega_2 Z_t}{W} \, dt \ \varepsilon \ \frac{1}{2} \, \mathbb{Z}$$

that is, if and only if

$$\int_0^1 \frac{\sqrt{3} \ \mathrm{i} B}{\sqrt{2}} \, t(1-t^2)^{-3/4} \, {}_2F_1\left(\frac{1}{12}, \frac{5}{12}; \ 1, \ t^2\right) dt \ \varepsilon \ \frac{1}{2} \, \mathbb{Z},$$

or letting $t^2 = s$,

$$\int_0^1 \frac{\sqrt{3} \ \mathrm{i} B}{2\sqrt{2}} \, (1-s)^{-3/4} \, {}_2F_1\left(\frac{1}{12}, \frac{5}{12}; \ 1; \ s \right) ds \ \varepsilon \ \frac{1}{2} \, \mathbb{Z} \ .$$

This is the type of integral that is frequently encountered in these examples. An excellent source for evaluating such integrals is Exton [40]. One finds

$$\int_0^1 (1-s)^{-3/4} \, {}_2F_1\left(\frac{1}{12}, \frac{5}{12}; \ 1, \ s\right) = \frac{\Gamma(1)\,\Gamma(1/4)\,\Gamma(3/4)}{\Gamma(5/6)\,\Gamma(7/6)} = \frac{6}{\sqrt{2}}$$

so that the periods will be integral if and only if

$$\frac{\sqrt{3} \ \mathrm{i} B}{2\sqrt{2}} \cdot \frac{6}{\sqrt{2}} \ \varepsilon \ \frac{1}{2} \, \mathbb{Z}$$

that is

$$B = \frac{-\mathrm{i} n}{3\sqrt{3}} \quad \text{for some } n \ \varepsilon \ \mathbb{Z} \ . \qquad \square$$

The reader will have little trouble in actually finding solutions
$(X(x):Y(x):1)$ to 1) and can easily verify with the help of Theorem V.3.1
that

$$\Lambda \int_{(0:1:0)}^{(X(x):Y(x):1)} \frac{dX}{Y} = \frac{\frac{-in}{3\sqrt{3}}}{x^2(x^2-1)^2}$$

for some $n \in \mathbb{Z}$. A solution with $n = \pm 1$ then generates the group of
solutions. In addition, for any solution $\mathscr{P} \in E^{gen}(\mathbb{C}(x))$ the normal
function $f = \int_{\mathcal{O}}^{\mathscr{P}} \frac{dX}{Y}$ satisfies the third order equation

$$0 = \frac{d^3f}{dx^3} + \frac{11x^2-3}{x(x^2-1)} \frac{d^2f}{dx^2} + \frac{29x^4 - 21\frac{5}{36}x^2 + \frac{8}{9}}{x^2(x^2-1)^2} \frac{df}{dx}$$

$$+ \frac{16x^2 - 6\frac{5}{18}}{x(x^2-1)^2} f$$

which has exponents:

$$0, \pm 1/3 \qquad \text{at} \quad 0 \quad \text{(see remark below)}$$

$$0, -1/4, -3/4 \quad \text{at} \quad \pm 1$$

$$2, 2, 4 \qquad\qquad \text{at} \quad \infty$$

and monodromy (taking $n = -1$, $c_1 = 0$, $c_2 = -\int_0^1 \frac{\omega_1 Z_t}{W} dt + 1/2$)

$$T_{\gamma_\infty} = \begin{pmatrix} 1 & 0 & 0 \\ 0 & 1 & 2 \\ 0 & 0 & 1 \end{pmatrix}$$

$$T_{\gamma_1} = \begin{pmatrix} 1 & 1 & 0 \\ 0 & 0 & -1 \\ 0 & 1 & 0 \end{pmatrix}$$

$$T_{\gamma_{-1}} = \begin{pmatrix} 1 & 1 & -1 \\ 0 & 1 & -2 \\ 0 & 1 & -1 \end{pmatrix}$$

$$T_{\gamma_0} = \begin{pmatrix} 1 & 1 & -1 \\ 0 & 0 & 1 \\ 0 & -1 & -1 \end{pmatrix} .$$

We remark that this third order differential equation satisfied by the normal functions can be identified in an interesting way -- the single-valued solutions at $x = 0$ happen to be

$$\{\frac{c}{x^2-1} \; {}_3F_2\binom{1/4,\ 3/4,\ 1}{5/6,\ 7/6}\ ;\ \frac{x^2}{x^2-1}) \quad c \in \mathbb{C}\}$$

where ${}_3F_2$ is the usual generalized hypergeometric function.

In addition we note that

$$\int_0^1 \frac{\omega_1 Z_t}{W} \, dt \qquad Z_t = \frac{Bt^2}{(1-t^2)^2} \quad B \in \mathbb{C}$$

can be explicitly evaluated as can the integrals of the first kind where

$$Z_t = \frac{At}{(1-t^2)^2}.$$

Example 2: Over the x-sphere $\mathbb{P}^1_{\mathbb{C}}$ consider the surface E given by:

2) $$E^{gen}: \ Y^2 = 4X^3 + 3(x^2-1)^{-1}X + (x^2-1)^{-2}$$

which has functional invariant $J = \dfrac{x^2-1}{x^2}$ and $\lambda^2 = 3(x^2-1)$. The K-equation annihilating the periods of $\dfrac{dX}{Y}$ for this model of E^{gen} is

$$\Lambda = \frac{d^2}{dx^2} + \frac{1}{x}\frac{d}{dx} + \frac{-\frac{1}{4}x^2 + \frac{41}{36}}{(x^2-1)^2}$$

which yields the following table:

x =	exponents	fiber	ord Z, $Z \in L_\Lambda^{para}(\mathcal{O}_L)$	ord Z, $Z \in L(\mathcal{O}_0)$
0	0,0	I_2	≥ -1	≥ -1
±1	1/3, 2/3	IV*	≥ -1	≥ -1
∞	-1/2, 1/2	I_0^*	≥ 2	≥ 3
elsewhere	0,1	good	≥ 0	≥ 0

;

thus

$$L_\Lambda^{para}(\theta_t) = \left\{ \frac{Ax+B}{x(x^2-1)} \quad A, B \in \mathbb{C} \right\}$$

and

$$L(\theta_0) = \left\{ \frac{B}{x(x^2-1)} \quad B \in \mathbb{C} \right\}.$$

We see that E is an elliptic K3 surface with Picard number $\rho = 20$ — one obvious solution is $(0:\pm(x^2-1)^{-1}:1)$. We will show that this solution generates the Mordell-Weil group $E^{gen}(\mathbb{C}(x))$ of 2) above.

We first analyze the monodromy representation of E. We take as a basis of solutions for Λ the functions

$$\omega_2(x) = \frac{\sqrt{2}\,\pi i}{\sqrt{3}} (1-x^2)^{1/4} \, {}_2F_1\left(\frac{1}{12}, \frac{5}{12}; 1; \frac{x^2}{x^2-1}\right)$$

$$\omega_1(x) = \text{etc.}$$

which analytically continue in the x-plane minus the slits pictured:

2 bis)

$$\begin{pmatrix} 0 & -1 \\ 1 & -1 \end{pmatrix} \qquad \begin{pmatrix} 1 & 2 \\ 0 & 1 \end{pmatrix} \qquad \begin{pmatrix} -1 & -1 \\ 1 & 0 \end{pmatrix}$$

to single valued functions with the indicated monodromy and which form a normalized basis giving the precise periods of dX/Y for 2) above. These functions $\omega_1(x)$, $\omega_2(x)$ are the pull-backs of the standard solutions $\omega_1(s)$, $\omega_2(s)$ of Proposition V.1.3 twisted by $\lambda = \sqrt{3}\, i \sqrt{1 - x^2}$. The picture is:

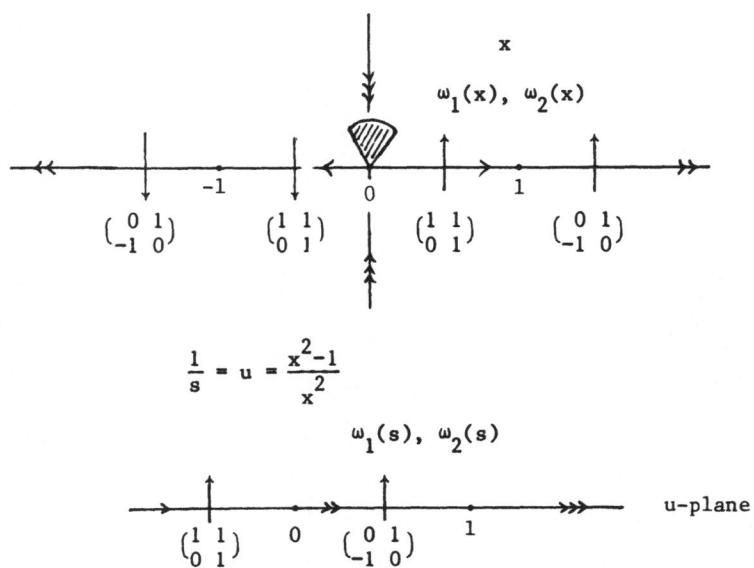

$$\frac{1}{s} = u = \frac{x^2-1}{x^2}$$

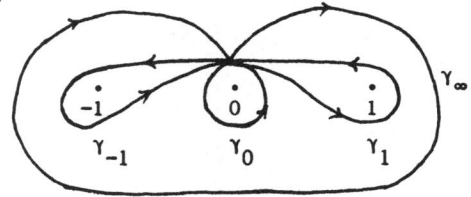

where we must in addition account for the sign changes imposed by λ and where the solutions are initially pulled back to a sector based at 0 which lies in the upper-half-plane. Taking the basis indicated below for $\pi_1(\mathbb{P}^1_\mathbb{C} - \{0, \pm 1, \infty\})$:

yields the monodromy representation

$$\gamma_{-1} \longmapsto M_{\gamma_{-1}} = \begin{pmatrix} 0 & -1 \\ 1 & -1 \end{pmatrix} = \begin{pmatrix} 1 & 1 \\ 1 & 0 \end{pmatrix}\begin{pmatrix} -1 & -1 \\ 1 & 0 \end{pmatrix}\begin{pmatrix} 1 & -1 \\ 0 & 1 \end{pmatrix}$$

$$\gamma_0 \longmapsto M_{\gamma_0} = \begin{pmatrix} 1 & 2 \\ 0 & 1 \end{pmatrix}$$

$$\gamma_1 \longmapsto M_{\gamma_1} = \begin{pmatrix} -1 & -1 \\ 1 & 0 \end{pmatrix}$$

$$\gamma_\infty \longmapsto M_{\gamma_\infty} = \begin{pmatrix} -1 & 0 \\ 0 & -1 \end{pmatrix}.$$

<u>Proposition V.4.3</u>: The global monodromy group $M \subset SL_2(\mathbb{Z})$ of E is the unique subgroup of index 2 in $SL_2(\mathbb{Z})$ generated by the squares of all the

elements in $SL_2(\mathbb{Z})$. It follows that the Mordell-Weil group $E^{gen}(\mathbb{C}(x))$ is torsion-free. (This is also clear from the fiber types.) $\quad\square$

Now we wish to consider the inhomogeneous equations of the form

$$\Lambda f = Z \qquad Z \in L_\Lambda^{para}(\mathcal{O}t) = \left\{ \frac{Ax+B}{x(x^2-1)} \quad A, B \in \mathbb{C} \right\}.$$

As always we have an expression for the normal function as

$$f = \left(\int_0^x \frac{-\omega_2 Z}{W} \, dx + c_1 \right)\omega_1(x) + \left(\int_0^x \frac{\omega_1 Z}{W} \, dx + c_2 \right)\omega_2(x)$$

which is single-valued on the slit x-plane described in 2 bis) above. We are interested in the periods or, as explained in section 2, the monodromy of the differential equation

$$\left(\frac{d}{dx} - \frac{d}{dx} \log Z \right) \cdot \Lambda$$

satisfied by the normal function f and ω_1, ω_2. The functions f, ω_1, ω_2 form a basis for the space of solutions of this third order equation with monodromy taking the form

$$\gamma_0 \mapsto T_{\gamma_0} = \begin{pmatrix} 1 & 0 & 2c_1 \\ 0 & 1 & 2 \\ 0 & 0 & 1 \end{pmatrix} \qquad \gamma_1 \mapsto T_{\gamma_1} = \begin{pmatrix} 1 & m_{\gamma_1} & n_{\gamma_1} \\ 0 & -1 & -1 \\ 0 & 1 & 0 \end{pmatrix}$$

$$\gamma_{-1} \mapsto T_{\gamma_{-1}} = \begin{pmatrix} 1 & m_{\gamma_{-1}} & n_{\gamma_{-1}} \\ 0 & 0 & -1 \\ 0 & 1 & -1 \end{pmatrix} \qquad \gamma_\infty \mapsto T_{\gamma_\infty} = \begin{pmatrix} 1 & m_{\gamma_\infty} & n_{\gamma_\infty} \\ 0 & -1 & 0 \\ 0 & 0 & -1 \end{pmatrix}$$

where $m_{\gamma_\infty} = m_{\gamma_1} + 2c_1 + m_{\gamma_{-1}} + n_{\gamma_{-1}}$ and $n_{\gamma_\infty} = n_{\gamma_1} - m_{\gamma_{-1}}$.

For which $Z \in L_\Lambda^{para}(\mathcal{O}\!\ell)$ will the periods be integral, so that f will be of the form $\int_\theta^{\mathscr{P}} \frac{dX}{Y}$ for some $\mathscr{P} \in E^{gen}(\mathbb{C}(x))$ corresponding to a section of $E/\mathbb{P}^1_\mathbb{C}$? The answer is given by:

<u>Proposition V.4.4</u>: $\Lambda f = Z$ for $Z \in L_\Lambda^{para}(\mathcal{O}\!\ell) = \{\frac{Ax+B}{x(x^2-1)} \quad A, B \in \mathbb{C}\}$ has integral periods if and only if $B = 0$ and $A = \frac{n}{6}$ for $n \in \mathbb{Z}$. Using Theorem V.3.1 we find that $\Lambda \int_{(0:1:0)}^{(0:(x^2-1)^{-1}:1)} \frac{dX}{Y} = \frac{-1/6}{(x^2-1)}$ and it follows that this solution generates the Mordell—Weil group $E^{gen}(\mathbb{C}(x))$.

<u>Proof</u>: It suffices to show that

$$\Lambda f = \frac{A}{(x^2-1)}$$

has integral periods if and only if $A = \frac{n}{6}$, $n \in \mathbb{Z}$. Given such an equation we want to know when we can choose c_1, c_2 mod $\mathbb{Z}$ so that $2c_1$, m_{γ_1}, n_{γ_1}, $m_{\gamma_{-1}}$, $n_{\gamma_{-1}} \in \mathbb{Z}$. We have

$$m_{\gamma_1} = -2 \int_0^1 \frac{-\omega_2 Z}{W} dx - 2c_1 + \int_0^1 \frac{\omega_1 Z}{W} dx + c_2$$

$$n_{\gamma_1} = - \int_0^1 \frac{-\omega_2 Z}{W} dx - c_1 - \int_0^1 \frac{\omega_1 Z}{W} dx - c_2$$

and

$$m_{\gamma_{-1}} = -2 \int_0^1 \frac{-\omega_2 Z}{W} dx - c_1 + \int_0^1 \frac{\omega_1 Z}{W} dx + c_2$$

$$n_{\gamma_{-1}} = \int_0^1 \frac{-\omega_2 Z}{W} dx - c_1 - 2 \int_0^1 \frac{\omega_1 Z}{W} dx - 2c_2$$

where for the latter we use that

$$W = \frac{-2\pi i}{3} \frac{1}{x}$$

so that

$$\frac{-\omega_2 Z}{W} dx = \frac{-\sqrt{3}\ A}{\sqrt{2}}\ x(1-x^2)^{-3/4}\ {}_2F_1\left(\frac{1}{12},\ \frac{5}{12};\ 1;\ \frac{x^2}{x^2-1}\right)dx$$

and

$$\frac{\omega_1 Z}{W} dx = \frac{\sqrt{3}\ A}{\sqrt{2}}\ x(1-x^2)^{-3/4}\ {}_2F_1\left(\frac{1}{12},\ \frac{5}{12};\ 1;\ \frac{x^2}{x^2-1}\right)\left(\frac{1}{\pi i}\ \log\ x - \frac{1}{2} - \frac{1}{2\pi i}\ \log(1-x^2)\right.$$

$$\left. + \text{ holomorphic in } x^2\right)$$

(here $-\pi/2 < \arg x < 3\pi/2$ in the branch of the log) thus giving

$$\int_0^{-1} \frac{-\omega_2 Z}{W}\ dx = \int_0^1 \frac{-\omega_2 Z}{W}\ dx$$

and

$$\int_0^{-1} \frac{\omega_1 Z}{W}\ dx = \int_0^1 \frac{\omega_1 Z}{W}\ dx - \int_0^1 \frac{-\omega_2 Z}{W}\ dx\ .$$

Now $m_{\gamma_{-1}} - m_{\gamma_1} = c_1$. So without loss of generality we can assume $c_1 = 0$. It is easy to see that the periods are integral if and only if

$$X_1 = \int_0^1 \frac{-\omega_2 Z}{W}\ dx\ \ \varepsilon\ \frac{1}{3}\ \mathbb{Z}$$

-- the result being achieved by setting

$$c_2 \equiv - \int_0^1 \frac{\omega_1 Z}{W}\ dx + \{0,\ 2/3,\ 1/3\ \text{mod } \mathbb{Z}$$

when $X_1 \equiv \{0,\ 1/3,\ 2/3\ \text{mod } \mathbb{Z}.$

We have that

$$X_1 = \int_0^1 \frac{-\sqrt{3}\ A}{\sqrt{2}}\ x(1-x^2)^{-3/4}\ {}_2F_1\left(\frac{1}{12},\ \frac{5}{12};\ 1;\ \frac{x^2}{x^2-1}\right)dx$$

$$= \int_0^{-\infty} \frac{\sqrt{3}\ A}{2\sqrt{2}}(1-s)^{-5/4}\ {}_2F_1\left(\frac{1}{12},\ \frac{5}{12};\ 1;\ s\right)ds,$$

via the substitution $s = \dfrac{x^2}{x^2-1}$,

$$= \frac{-\sqrt{3}\ A}{2\sqrt{2}} \int_0^\infty (s+1)^{-5/4} \,\, {}_2F_1\left(\tfrac{1}{12},\ \tfrac{5}{12};\ 1;\ -s\right) ds$$

$$= \frac{-\sqrt{3}\ A}{2\sqrt{2}} \ \frac{\Gamma(1)\,\Gamma(1/3)\,\Gamma(2/3)}{\Gamma(5/4)\,\Gamma(3/4)} \,\, {}_2F_1\left(\tfrac{1}{3},\ \tfrac{2}{3};\ \tfrac{3}{4};\ 0\right)$$

(see Gradshteyn and Ryzhik [39] pg. 849 or Exton [40])

$$= \frac{-\sqrt{3}\ A}{2\sqrt{2}} \cdot \frac{\pi\ \csc(\pi/3)}{1/4\ \pi\ \csc(\pi/4)} = -2\ A$$

and the result follows: $-2A$ is in $\tfrac{1}{3}\,\mathbb{Z}$ iff $A = \tfrac{n}{6}$, $n \in \mathbb{Z}$. $\quad\square$

Finally, any normal function $f = \int_\theta^\varphi \dfrac{dX}{Y}$ for $\varphi \in E^{gen}(\mathbb{C}(x))$ will

satisfy the third order equation

$$0 = \frac{d^3 f}{dx^2} + \frac{3x^2-1}{x(x^2-1)}\,\frac{d^2 f}{dx^2} + \frac{\tfrac{3}{4}x^4 + \tfrac{41}{36}x^2 - 1}{(x^2-1)^2 x^2}\,\frac{df}{dx} + \frac{-\tfrac{16}{9}x}{(x^2-1)^3}\,f$$

which has monodromy (taking $c_1 = 0$, $n = -1$, $c_2 = -\int_0^1 \dfrac{\omega_1 Z}{W}\,dx + \tfrac{2}{3}$)

$$\gamma_0 \rightarrow T_{\gamma_0} = \begin{pmatrix} 1 & 0 & 0 \\ 0 & 1 & 2 \\ 0 & 0 & 1 \end{pmatrix}$$

$$\gamma_1 \rightarrow T_{\gamma_1} = \begin{pmatrix} 1 & 0 & -1 \\ 0 & -1 & -1 \\ 0 & 1 & 0 \end{pmatrix}$$

$$\gamma_{-1} \rightarrow T_{\gamma_{-1}} = \begin{pmatrix} 1 & 0 & -1 \\ 0 & 0 & -1 \\ 0 & 1 & -1 \end{pmatrix}$$

$$\gamma_\infty \rightarrow T_{\gamma_\infty} = \begin{pmatrix} 1 & -1 & -1 \\ 0 & -1 & 0 \\ 0 & 0 & -1 \end{pmatrix}$$

and which can be related to a generalized hypergeometric equation ${}_3F_2$ as

in the previous example.

Example 3: This set of examples was originally discussed by Sasai (Sasai [24]). We will determine the Picard numbers using quite a different method coming out of the results in Part IV. Our base curve X will be the sphere and we will denote the parameter by x (i.e. X is the x-sphere). A model for E over $K(X) = \mathbb{C}(x)$ can be given as

1) $$Y^2 = 4X^3 - \frac{27}{x^{12k}}X - \frac{27}{x^{12k}} \qquad k \geq 1 \ .$$

Let ζ be the primitive $12k^{\text{th}}$ root of unity $e^{(2\pi i/12k)}$. One sees immediately that 1) defines an elliptic surface over $X - \{0,\infty\}$ whose singular fibers are of type I_1 at ζ^i $i = 0, \ldots, 12k-1$. The fibers over 0 and ∞ are in fact good.

In the model 1) $J = \frac{1}{1-x^{12k}}$ and $\lambda = 1$ so the K-equation $\Lambda = \Lambda_{(\ ,1)}$ is the differential equation annihilating the periods of dX/Y. The exact period functions can be determined as in the previous section.

Proposition V.4.8: For the model 1) above the K-equation annihilating the periods of dX/Y is $\Lambda = \Lambda_{(J,1)}$ which is

$$\frac{d^2f}{dx^2} + \frac{(1 - x^{12k}) - 12k}{x(1 - x^{12k})}\frac{df}{dx} + \frac{(27k^2 + 4k^2x^{12k})}{x^2(1 - x^{12k})}f \ .$$

Proof: If J were used as the parameter in forming Λ the equation would be

$$\frac{d^2f}{dJ^2} + \frac{1}{J}\frac{df}{dJ} + \frac{\frac{31}{144}J - \frac{1}{36}}{J^2(J-1)^2}f$$

which is

$$\frac{d^2f}{dJ^2} + (1 - x^{12k})\frac{df}{dJ} + \frac{(1 - x^{12k})^3(\frac{3}{16} + \frac{1}{36})x^{12k}}{x^{24k}}f \ .$$

We now change parameter to x from J using the formula on page 7 and get $\Lambda_{(J,1)}$ in terms of x to be:

$$\frac{d^2f}{dx^2} + \frac{(1-x^{12k}) - 12k}{x(1-x^{12k})}\frac{df}{dx} + \frac{(27k^2 + 4k^2x^{12k})}{x^2(1-x^{12k})}\, f \ . \qquad \square$$

This equation is regular except at 0, ∞, ζ^i $i = 0,\ldots,12k-1$. We shall investigate its local behavior. First at $x = 0$ we have

$$\frac{d^2f}{dx^2} + \left(\frac{1-12k}{x} + (-12k)x^{12k-1} + \ldots\right)\frac{df}{dx} + \left(\frac{27k^2}{x^2} + 31k^2x^{12k-2} + \ldots\right)f$$

The indicial equation is $v(v-1) + (1-12k)v + 27k^2 = v^2 - 12kv + 27k^2$, so the exponents are $3k$, $9k$ and $x = 0$ is a cosingular point. At ∞, letting $t = 1/x$, the equation is

$$\frac{d^2f}{dt^2} + \left(\frac{1}{t} + (-12k)t^{12k-1} + \ldots\right)\frac{df}{dt} + \left(\frac{-4k^2}{t^2} + (-31k^2)t^{12k-2} + \ldots\right)f \ .$$

The indicial equation is $v(v-1) + v - 4k^2 = v^2 - 4k^2$, so the exponents are $-2k$, $2k$ and $x = \infty$ is a cosingular point. Finally at $x = \zeta^i$ $i = 0 \ldots 12k-1$ we have

$$\frac{d^2f}{dx^2} + \left(\frac{1}{x-\zeta^i} + \left(\frac{1-12k}{2}\right)(\zeta^i)^{-1} + \ldots \quad \right)\frac{df}{dx} + \left(\frac{0}{(x-\zeta^i)^2} + \frac{-\frac{31}{12}k(\zeta^i)^{-1}}{(x-\zeta^i)} + \ldots \right)f.$$

Here the exponents are $(0,0)$ and the points $x = \zeta^i$ are true singularities. Note that all of this agrees with our previous local calculations in Part II.

The basic elliptic surface corresponding to our model 1) has geometric genus $p_g = k-1$, irregularity $q = 0$, and $\chi(\mathcal{O}_E) = k$. The monodromy representation is actually easy to compute (see Sasai [24] or Stiller [30]). The global monodromy group is $SL_2(\mathbb{Z})$ and $E^{gen}(K(X))$, the group of $K(X)$-rational points on the generic fiber, is torsion free because all the singular fibers are of type I_1.

We wish to prove:

Theorem V.4.9: Let r denote the rank of the Mordell-Weil group $E^{gen}(\mathbb{C}(x))$ of E^{gen}: $Y^2 = 4X^3 - \dfrac{27}{x^{12k}} X - \dfrac{27}{x^{12k}}$ and let ρ be the Picard number of the associated elliptic surface, then

$$r = \sum_{\substack{d \mid 12k \\ d \geq 1 \\ d \text{ admissible}}} \phi(d)$$

where ϕ is Euler's function, so $\phi(d)$ is the number of positive integers $\leq d$ relatively prime to d, and d is admissible if no primitive dth root of unity has argument between $\pi/3$ and $\pi/2$ inclusive. It follows that

$$\rho = 2 + \sum_{\substack{d \mid 12k \\ d \geq 1 \\ d \text{ admissible}}} \phi(d) \qquad . \qquad \qquad \Box$$

It can be shown that there are ony eleven admissible numbers: 1,2,3,7,8,10,12,15,18,20, and 42. Thus the formula becomes

$$r = 8 + \sum_{\substack{d \mid 12k \\ d \in \{7,8,10,15,18,20,42\}}}$$

hence $r \leq 56$ with equality if and only if $k \equiv 0 \bmod 210$.

Previously, the best known estimate for r was $r \leq 10k - 2$ (see Stiller [29]).

Before considering the proof of Theorem V.4.9, we need to study the periods and the spaces $L_\Lambda^{para}(\mathcal{O}\!l)$, $L(\mathcal{O}\!l_0)$.

Recall that given a section of E over X or equivalently a $K(X)$-rational point $(X(x), Y(x))$ on the generic fiber of E/X (see model 1) above), we can produce an element of $K(X)$ via the map

$$E^{gen}(K(X)) \longrightarrow K(X)$$

$$\mathcal{P} = (X(x), Y(x)) \longrightarrow \Lambda_{(J,1)} \int_{\mathcal{O}}^{\mathcal{P}} \frac{dX}{Y}$$

$$= \Lambda_{(J,1)} \int_{(0:1:0)}^{(X(x):Y(x):1)} \frac{dX}{Y}$$

and this map gives an injective homomorphism from the group of sections to the additive group $K(X)$. This map has been explicitly computed (Part V, §3 above). For our purposes it is enough to observe that the image lies in $L_\Lambda^{para}(\mathcal{O}\!l)$ where

$$\mathcal{O}\!l = \sum_{i=0}^{12k-1} (1)\zeta^i + (-3k+1)0 + (2k-3)\infty$$

(see Theorem III.2.10). We have

$$\deg \mathcal{O}\!l = 11k - 2$$

and

$$\dim L(\mathcal{O}\!l) = 11k - 1$$

A typical element of $L(\mathcal{O}\!\ell)$ can be written

$$Z = \frac{x^{3k-1}(\text{polynomial of degree} \leq 11k-2)}{(1 - x^{12k})} \; .$$

In order for Z to be in $L_{\Lambda}^{\text{para}}(\mathcal{O}\!\ell) \subset L(\mathcal{O}\!\ell)$ we need to check the residue conditions at the cosingular points 0, ∞. Near $x = 0$ this is the condition that

$$\frac{\omega Z}{W} \, dx \qquad W \text{ the Wronskian}$$

have zero residue for any solution $\Lambda\omega = 0$ at 0. Now up to a constant

$$W = c \, \frac{dJ/dx}{J} = c \, \frac{12kx^{12k-1}}{(1-x^{12k})} \qquad c \neq 0$$

and so we must consider the differential

$$\omega \Big(\frac{\text{polynomial of degree} \leq 11k-2}{(c12k)x^{9k}} \Big) dx$$

or

$$\omega \Big(\frac{p(x)}{c12kx^{9k}} \Big) dx \qquad \deg p(x) \leq 11k-2 \; .$$

Near $x = 0$ the differential equation can be seen to have a basis of solutions (not necessarily a K-basis) of the form

$$x^{3k} (1 + cx^{12k} + \ldots)$$

$$x^{9k} (\text{holomorphic}) \; .$$

As a result, if

$$p(x) = a_{11k-2} \, x^{11k-2} + \ldots + a_0$$

we must have $a_{6k-1} = 0$ to assure zero residues at $x = 0$. A similar calculation at ∞ shows we must have $a_{7k-1} = 0$ to assure zero residues at $x = \infty$.

Thus a typical element of $L_\Lambda^{para}(\mathcal{R}) \subset L(\mathcal{R})$ is of the form

$$\frac{x^{3k-1}(a_{11k-2}x^{11k-2}+\ldots+a_{7k}x^{7k}+a_{7k-2}x^{7k-2}+\ldots+a_{6k}x^{6k}+a_{6k-2}x^{6k-2}+\ldots+a_0)}{(1-x^{12k})}$$

and

$$\dim L_\Lambda^{para}(\mathcal{R}) = 11k - 3 .$$

The space of elements of the "first kind" is $L(\mathcal{R}_0)$ where

$$\mathcal{R}_0 = \sum_{i=0}^{12-1} (1)\zeta^i + (-9k+1)0 + (-2k-3)\infty$$

(see page 85). And a typical element $Z \in L(\mathcal{R}_0)$ is of the form

$$\frac{x^{3k-1}(a_{7k-2}x^{7k-2} + \ldots + a_{6k}x^{6k})}{(1-x^{12k})}$$

that is

$$\frac{x^{9k-1}(\text{polynomial of degree} \leq k-2)}{(1-x^{12k})} .$$

We can now give the proof of Theorem V.4.9.

Proof: Consider the automorphism $\phi: E \rightarrow E$ induced by the map $x \rightarrow \zeta x$, $\zeta = e^{2\pi i/12k}$ on the base curve $X \cong \mathbb{P}^1_{\mathbb{C}}$. The resulting automorphism on cohomology, in particular on $H^2(E, \mathbb{C})$, preserves the Leray filtration (recall that the Leray spectral sequence for $\pi: E \rightarrow X$ and the constant sheaf $\mathbb{C}$ on E degenerates at E_2^{pq}) and the Hodge decomposition.

We shall analyze this automorphism in terms of the action on $L_\Lambda^{para}(\mathcal{O}l)$. This action is easily seen to be

$$L^{para}(\mathcal{O}l) \longrightarrow L_\Lambda^{para}(\mathcal{O}l)$$

$$Z(x) \longrightarrow Z(\zeta x)\zeta^2 \qquad \zeta = e^{2\pi i/12k}$$

(the extra factors of ζ are because one should work with $Z(dx)^2$ in order to get a parameter invariant description).

One could trace back through our description of the parabolic cohomology class determined by an equation of the second kind to see that this corresponds to the action of our automorphism on the $2,0$ and $1,1$ parts of $E_2^{1,1} = H^1(X, R^1\pi_*C)$, but the reader can see this in a more direct fashion.

Suppose we consider an inhomogeneous equation

$$\Lambda_x f(x) = Z(x) \qquad Z \in L_\Lambda^{para}(\mathcal{O}l)$$

where we have subscripted Λ with an x to indicate the parameter used in forming Λ

$$\Lambda_x = \frac{d^2}{dx^2} + P(x)\frac{d}{dx} + Q(x) .$$

Denote by $\Lambda_{\zeta x}$ the differential equation

$$\Lambda_{\zeta x} = \frac{d^2}{d(\zeta x)^2} + P(\zeta x)\,\frac{d}{d(\zeta x)} + Q(\zeta x)$$

and observe that in our case

$$\Lambda_{\zeta x} = \zeta^{-2}\Lambda_x \; .$$

Now consider the function $f(\zeta x)$. Because $\omega_1(\zeta x)$, $\omega_2(\zeta x)$ are linear combinations of $\omega_1(x)$, $\omega_2(x)$, we have that

$$\Lambda_x f(\zeta x) = \zeta^2 \Lambda_{\zeta x} f(\zeta x) = \zeta^2 Z(\zeta x) \; .$$

This describes the action of our automorphism. For example, consider the action on $E^{gen}(\mathbb{C}(x))$ — if $(X(x):Y(x):1)$ is a solution to 1) above then $(X(\zeta x):Y(\zeta x):1)$ is also and we have

$$\Lambda_x \int_{(0:1:0)}^{(X(x):Y(x):1)} \frac{dX}{Y} = Z(x)$$

and

$$\Lambda_x \int_{(0:1:0)}^{(X(\zeta x):Y(\zeta x):1)} \frac{dX}{Y} = \zeta^2 Z(\zeta x)$$

because if we let

$$f(x) = \int_{(0:1:0)}^{(X(x):Y(x):1)} \frac{dX}{Y}$$

be the normal function, then

$$\int_{(0:1:0)}^{(X(\zeta x):Y(\zeta x):1)} \frac{dX}{Y} = f(\zeta x) \; .$$

Thus we have the diagram:

$$(X(x):Y(x):1) \quad \varepsilon \quad E^{gen}(\mathbb{C}(x)) \longrightarrow \mathbb{C}(x) \ni Z(x)$$

$$(X(\zeta x):Y(\zeta x):1) \quad \varepsilon \quad E^{gen}(\mathbb{C}(x)) \longrightarrow \mathbb{C}(x) \ni Z(\zeta x)\zeta^2$$

where the vertical left hand arrow is the map induced on sections by our automorphism and the horizontal arrows are Manin's map (see Part III, §1).

Also consider the action on two-forms. Recall that the space of generalized automorphic forms of the first kind can be identified with $H^0(E, \Omega_E^2)$ via

$$g \rightarrow g(z)dz \wedge dt \qquad z \varepsilon \hslash, \quad t \varepsilon \mathbb{C}$$

(see Theorem III.5.7). Using our explicit formula for g, we see that this is essentially

$$\frac{-\omega_2 Z}{W} \quad dx \wedge dt .$$

But remember that the Wronskian W depends on the choice of derivation $\frac{d}{dx}$. A parameter free version would be

$$\frac{-\omega_2(x)Z(x)(dx)^2}{Wdx} \wedge dt .$$

Now both $\omega_2(x)$ and $W(x)dx$ are invariant under $x \rightarrow \zeta x$, so we see that the action on 2-forms corresponds to the action on $L(\mathcal{O}_0) \subset L_\Lambda^{para}(\mathcal{O})$.

Now consider the eigenspaces for the action of our automorphism on $L_\Lambda^{para}(\mathcal{O})$. All the eigenvalues are $12k^{th}$ roots of unity and we see that the eigenspaces are all of the form

$$\mathbb{C} \cdot \frac{X^i}{1-X^{12k}}$$

which has eigenvalue ζ^{i+2}. From this we get the important fact that the eigenvalues of our automorphism on $H^1(X,R^1\pi_*\mathbb{C})$ occur with multiplicity one -- all the eigenspaces are one dimensional. Now our automorphism preserves type so the $(2,0)$, $(1,1)$, and $(0,2)$ pieces are sums of distinct eigenspaces. The relevant picture is:

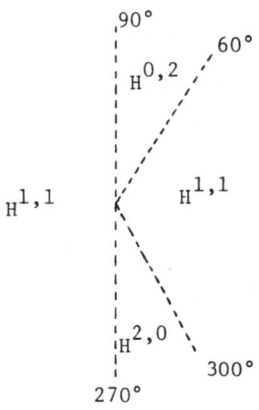

This gives us the Hodge decomposition of $L_\Lambda^{para}(\mathcal{A})$ into the $(2,0)$ part (namely $L(\mathcal{A}_0)$) and the complementary $(1,1)$ part. As our automorphism preserves rational cohomology it can be decomposed over $\mathbb{Q}$ into invariant subspaces over all primitive d^{th} roots of unity for some $d \mid 12k$. Thus if the eigenspaces corresponding to all primitive d^{th} roots fall in $H^{1,1}$, we have a rational subspace. The result follows. □

Example 4: We consider over the sphere $\mathbb{P}^1_\mathbb{C}$ the elliptic surfaces E given by

3) $$E^{gen}: Y^2 = 4X^3 - 3x^{3k}X - x^{5k} \qquad k > 0$$

which have functional invariant $J = \dfrac{1}{1-x^k}$ and $\lambda^2 = \dfrac{3}{x^{2k}}$. The K-equation annihilating the periods of $\dfrac{dX}{Y}$ for this model of E^{gen} is

$$\Lambda = \frac{d^2}{dx^2} + \frac{(2k+1)x^k - (k+1)}{x(x^k-1)} \frac{d}{dx} + \frac{\frac{35}{36} k^2 x^k - \frac{27}{144} k^2}{x^2(x^k-1)} .$$

The resulting surface is an elliptic K3 surface for k = 5, 9, 10, 11, 13,

14, 15, 16, 18, 19, 20 and 24.

Proposition V.4.7: We have:

k	fiber type at 0	fiber type at ∞	fiber over kth roots 1	r	ρ
5	III*	II*	I_1	1	18
9	III*	I_0^*	I_1	3	16
10	I_0^*	IV*	I_1	6	18
11	III	II*	I_1	1	12
13	III*	II	I_1	1	10
14	I_0^*	IV	I_1	8	16
15	III	I_0^*	I_1	11	18
16	good	IV*	I_1	6	14
18	I_0^*	good	I_1	10	16
19	III	II	I_1	1	4
20	good	IV	I_1	14	18
24	good	good	I_1	12	14

where r is the rank of the Mordell-Weil group $E^{gen}(\mathbb{C}(x))$ and ρ is the

Picard number.

Proof: The Picard number for arbitrary k is computed in Stiller [43]

where complete proofs appear.   □

Example 5: Consider over the sphere the surface whose generic fiber is

given by

$$Y^2 = 4X^3 - 3\,\frac{1}{(x^2-1)^2}\,X - \frac{(x^2-2)}{x^2(x^2-1)^3}$$

which has functional invariant $J = \dfrac{x^4}{4(x^2-1)}$ and $\lambda^2 = \dfrac{3x\,(x^2-1)}{(x^2-x)}$. The

K-equation annihilating the periods of $\dfrac{dX}{Y}$ for this model is

$$\Lambda = \frac{d^2}{dx^2} + \frac{-1}{x}\frac{d}{dx} + \frac{x^4 - 5/9x^2 + 5/9}{x^2(x^2-1)^2}$$

which yields the following data:

$x =$	exponents	fiber	ord Z, $Z \in L_\Lambda^{para}(\mathcal{O}l)$	ord Z, $Z \in L(\mathcal{O}l_0)$
0	1/3, 5/3	IV*	≥ -1	≥ 0
+1	1/2, 1/2	I_1^*	≥ -1	≥ -1
-1	1/2, 1/2	I_1^*	≥ -1	≥ -1
∞	-1 -1	I_2	≥ 2	≥ 2

We see that our surface is an elliptic K3 surface with $\rho = 19$ or 20
($r = 0$ or 1) with

$$L_\Lambda^{para}(\mathcal{O}l) = \{\frac{Ax+B}{x(x^2-1)}\ ;\ A,\ B \in \mathbb{C}\}$$

$$L(\mathcal{O}l_0) = \{\frac{A}{(x^2-1)}\ ;\ A \in \mathbb{C}\}\ .$$

<u>Proposition V.4.10:</u> $\Lambda f = Z$ for $Z \in L_\Lambda^{para}(\mathcal{O}l) = \{\dfrac{Ax+B}{x(x^2-1)}\ ;\ A,\ B \in \mathbb{C}\}$ has

integral periods if and only if $A = 0$ and $B = \dfrac{21}{3\sqrt6}\,n$ for $n \in \mathbb{Z}$. It follows that the Mordell-Weil group $E^{gen}(\mathbb{C}(x)) = \mathbf{Z}$ and that the Picard number of the surface above is 20.

As usual we begin by considering the monodromy representation of Λ. To do this we change parameter to $t = 1/x$ and consider the slit t-plane

together with the basis of solutions for Λ in the sector based at $t = 0$ $(x = \infty)$ along the positive real axis as shown

$$\omega_2(t) = \frac{\sqrt{2}\pi}{\sqrt{3}} t^{-1}(1-t^2)^{1/2} \,_2F_1(1/6, 5/6; 1; t^2)$$

$$\omega_1(t) = \omega_2(t)\left(\frac{1}{\pi i} \log t + \frac{1}{2\pi i} \log \frac{4}{1728} + \text{holomorphic vanishing in } t^2\right)$$

Here we assume $-\pi/2 < \arg t < 3\pi/2$ in our branch of the log and that $\sqrt{1-t^2}$ is normalized to take value 1 at $t = 0$. It is relatively straightforward to verify that these functions form a normalized basis -- that is, they give the precise periods of dX/Y. (Note that $\,_2F_1(1/6,5/6;1;t^2) = \,_2F_1(1/12,5/12;1;4t^2(1-t^2))$; see Bateman [1]). For the illustrated basis of π_1,

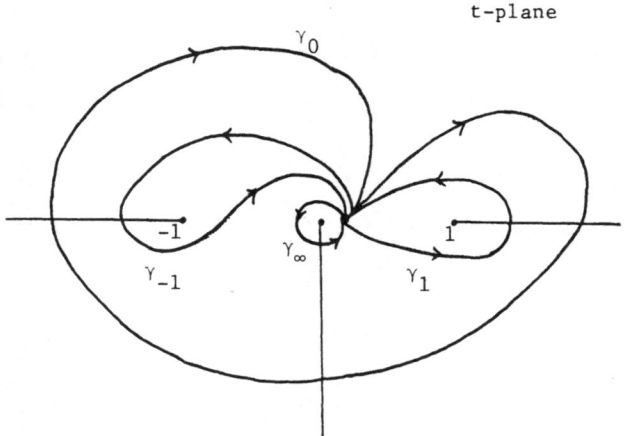

t-plane

one can show that the monodromy representation of Λ with respect to ω_1, ω_2 is given by

$$\gamma_\infty \longrightarrow M_{\gamma_\infty} = \begin{pmatrix} 1 & 2 \\ 0 & 1 \end{pmatrix}$$

$$\gamma_1 \longrightarrow M_{\gamma_1} = \begin{pmatrix} -1 & 0 \\ 1 & -1 \end{pmatrix}$$

$$\gamma_{-1} \longrightarrow M_{\gamma_{-1}} = \begin{pmatrix} 0 & -1 \\ 1 & -2 \end{pmatrix}$$

$$\gamma_0 \longrightarrow M_{\gamma_0} = \begin{pmatrix} 0 & -1 \\ 1 & -1 \end{pmatrix} .$$

Notice that this last is conjugate to

$$\begin{pmatrix} -1 & -1 \\ 1 & 0 \end{pmatrix} .$$

One sees that the global monodromy group is $M = SL_2(\mathbb{Z})$. Now consider an inhomogeneous equation of the form

$$\Lambda f = Z \qquad\qquad Z \in L_\Lambda^{para}(\mathcal{O}_t)$$

The periods (expressed in terms of the monodromy of the third order

equation $\left(\dfrac{d}{dx} - \dfrac{d}{dx}\log Z\right) \circ \Lambda$ satisfied by the "normal function") take

the form

$$T_{\gamma_\infty} = \begin{pmatrix} 1 & 0 & 2c_1 \\ 0 & 1 & 2 \\ 0 & 0 & 1 \end{pmatrix}$$

$$T_{\gamma_1} = \begin{pmatrix} 1 & m_{\gamma_1} & n_{\gamma_1} \\ 0 & -1 & 0 \\ 0 & 1 & -1 \end{pmatrix}$$

$$T_{\gamma_{-1}} = \begin{pmatrix} 1 & m_{\gamma_{-1}} & n_{\gamma_{-1}} \\ 0 & 0 & -1 \\ 0 & 1 & -2 \end{pmatrix}$$

$$T_{\gamma_0} = \begin{pmatrix} 1 & m_{\gamma_0} & n_{\gamma_0} \\ 0 & 0 & -1 \\ 0 & 1 & -1 \end{pmatrix}$$

with respect to the basis f, ω_1, ω_2 where

$$f(t) = \left(\int_0^t \frac{-\omega_2 Z}{W}\, dt + c_1\right)\omega_1(t) + \left(\int_0^t \frac{\omega_1 Z}{W}\, dt + c_2\right)\omega_2(t)$$

which is single-valued on the t-plane slit as above. As usual we have

concrete expressions for the various terms above:

$$W = \frac{2\pi i}{3} \cdot \frac{1}{t^3}$$

$$\frac{-\omega_2 Z}{W}\, dt = \frac{\sqrt{3}i}{\sqrt{2}}\, (1-t^2)^{-1/2}(A+Bt)\; {}_2F_1(1/6, 5/6; 1; t^2)\, dt$$

$$\frac{\omega_1 Z}{W} \, dt = -\frac{\sqrt{3i}}{\sqrt{2}} (1-t^2)^{-1/2} (A+Bt) \, {}_2F_1(1/6,5/6;1;t^2) \left(\frac{1}{\pi i} \log t + \dots\right) dt$$

(note that Z becomes $Z\left(\frac{dx}{dt}\right)^2$ under our change of parameters)

$$m_{\gamma_1} = -2 \int_0^1 \frac{-\omega_2 Z}{W} \, dt - 2c_1 + \int_0^1 \frac{\omega_1 Z}{W} \, dt + c_2$$

$$n_{\gamma_1} = -2 \int_0^1 \frac{\omega_1 Z}{W} \, dt - 2c_2$$

$$m_{\gamma_{-1}} = -\int_0^{-1} \frac{-\omega_2 Z}{W} \, dt - c_1 + \int_0^{-1} \frac{\omega_1 Z}{W} \, dt + c_2$$

$$n_{\gamma_{-1}} = \int_0^{-1} \frac{-\omega_2 Z}{W} \, dt - c_1 - 3 \int_0^{-1} \frac{\omega_1 Z}{W} \, dt - 3c_2$$

(note that all the integrals converge)

$$m_{\gamma_0} = -n_{\gamma_1} + 2c_1 + 2m_{\gamma_{-1}} + n_{\gamma_{-1}}$$

$$n_{\gamma_0} = m_{\gamma_1} + n_{\gamma_1} - m_{\gamma_{-1}} \, .$$

Now to prove Proposition V.4.10 above it suffices to show that

$\Lambda f = \dfrac{B}{x(x^2-1)}$ has integral periods if and only if $B = \dfrac{2i}{3\sqrt{6}} \, n$, $n \in \mathbb{Z}$. In

this case

$$\int_0^{-1} \frac{-\omega_2 Z}{W} \, dt = \int_0^1 \frac{\omega_2 Z}{W} \, dt$$

and

$$\int_0^{-1} \frac{\omega_1 Z}{W} \, dt = \int_0^1 \frac{\omega_1 Z}{W} \, dt - \int_0^1 \frac{-\omega_2 Z}{W} \, dt$$

so that

$$m_{\gamma_{-1}} = -2 \int_0^1 \frac{-\omega_2 Z}{W} \, dt - c_1 + \int_0^1 \frac{\omega_1 Z}{W} \, dt + c_2$$

$$n_{\gamma_{-1}} = 2 \int_0^1 \frac{-\omega_2 Z}{W} \, dt - c_1 - 3 \int_0^1 \frac{\omega_1 Z}{W} \, dt - 3c_2 \ .$$

We see that $m_{\gamma_{-1}} - m_{\gamma_1} = c_1$, so that we can get integer periods if and only if

$$2 \int_0^1 \frac{-\omega_2 Z}{W} \, dt \ \epsilon \ 1/2 \ \mathbb{Z}$$

by setting $c_2 = -\int_0^1 \frac{\omega_1 Z}{W} \, dt + 1/2$ modulo $\mathbb{Z}$ if $2 \int_0^1 \frac{-\omega_2 Z}{W} \, dt \not\equiv 0$ modulo $\mathbb{Z}$

and $c_2 = -\int_0^1 \frac{\omega_1 Z}{W} \, dt$ modulo $\mathbb{Z}$ otherwise and choosing $c_1 \equiv 0$ mod $\mathbb{Z}$. Thus we get integer periods exactly when

$$2\sqrt{6} \ \text{Bi} \int_0^1 t(1-t^2)^{1/2} \ _2F_1(1/6,5/6;1;t^2) dt \ \epsilon \ \mathbb{Z}$$

or

$$\sqrt{6} \ \text{Bi} \int_0^1 (1-s)^{-1/2} \ _2F_1(1/6,5/6;1;s) ds \ \epsilon \ \mathbb{Z}$$

$$\sqrt{6} \ \text{Bi} \ \frac{\Gamma(1)\Gamma(1/2)\Gamma(1/2)}{\Gamma(4/3)\Gamma(2/3)} = \frac{3\sqrt{6}}{2} \ \text{iB} \ \epsilon \ \mathbb{Z}$$

which establishes the result .   □

The normal functions $f = \int_{\partial}^{\rho} \frac{dX}{Y}$, $\rho \epsilon \ E^{\text{gen}} \ (\mathbb{C}(x))$ satisfy the third order equation

$$\frac{d^3}{dx^3} + \frac{2x}{(x^2-1)} \frac{d^2}{dx^2} + \frac{-x^4 + 13/9 \ x^2 + 5/9}{x^2(x^2-1)^2} \frac{d}{dx} + \frac{x^6 - 22/9 \ x^4 - 10/9x^2 + 5/9}{x^3(x^2-1)^3}$$

which has exponents $1/3$, $5/3$, 1 at 0; $1/2$, $1/2$, 1 at ±1; -1, -1, 1 at ∞ and which has monodromy $\left(c_1 = 0, \ B = -\frac{2i}{3\sqrt{6}} \right)$

$$\gamma_\infty \quad \to T_{\gamma_\infty} = \begin{pmatrix} 1 & 0 & 0 \\ 0 & 1 & 2 \\ 0 & 0 & 1 \end{pmatrix}$$

$$\gamma_1 \quad \to T_{\gamma_{-1}} = \begin{pmatrix} 1 & 0 & -1 \\ 0 & -1 & 0 \\ 0 & 1 & -1 \end{pmatrix}$$

$$\gamma_{-1} \quad \to T_{\gamma_{-1}} = \begin{pmatrix} 1 & 0 & -1 \\ 0 & 0 & -1 \\ 0 & 1 & -2 \end{pmatrix}$$

$$\gamma_0 \quad \to T_{\gamma_0} = \begin{pmatrix} 1 & 0 & -1 \\ 0 & 0 & -1 \\ 0 & 1 & -1 \end{pmatrix} .$$

Moreover this third order equation can be more explicitly identified — its space of single-valued solutions at $x = 0$ is all multiples of

$$_3F_2\binom{1/2,1/2,1}{2/3,4/3} ; \frac{x^2}{x^2-1})$$

where $_3F_2$ is the usual generalized hypergeometric function (see Bateman [1]).

Example 6: Again our base curve X will be the sphere $\mathbb{P}^1_{\mathbb{C}}$ and we will denote the parameter by x (i.e. $X \simeq \mathbb{P}^1_{\mathbb{C}}$ is the x-sphere). We consider the basic elliptic surfaces E_k, $k \in \mathbb{Z}$, $k > 0$ given in terms of a model for the generic fiber E_k^{gen} over $\mathbb{C}(x)$ by

$$Y^2 = 4X^3 - 3x^{12k+3}(x^{4k+1} - 1/4)X - x^{20k+5}(x^{4k+1} + 1/8) .$$

These surfaces have singular fibers of type I_1 at the $4k+1^{st}$ roots of unity, I_{8k+2} at ∞ and III^* at 0. The geometric genus is $p_g = k$ and one can show:

Proposition V.4.11: The surface E_k^{gen} has Picard number

$$\rho = \begin{cases} 8k + 10 & k \equiv 0,1 \mod 3 \\ 8k + 12 & k \equiv 2 \mod 3 \end{cases}$$

or equivalently the rank of the Mordell-Weil group $E_k^{gen}(\mathbb{C}(x))$ is

$$r = \begin{cases} 0 & k \equiv 0,1 \mod 3 \\ 2 & k \equiv 2 \mod 3 \end{cases}$$

Proof: See Stiller [43].   □

Example 7: We again consider surfaces E_k, $k \in \mathbb{Z}$, $k > 0$ given in terms of a model for the generic fiber over $\mathbb{C}(x)$ by

$$Y^2 = 4X^3 - 3x^{4k}X + x^{5k}(x^k-2) .$$

If we write $k = 6\ell + r$ with $0 < r \leq 6$, then the geometric genus is $p_g = \ell$ and the surface has a singular fiber of type I_k over ∞, singular fibers of type I_1 over the kth roots of unity, and a singular fiber of type II^* (resp. IV^*, I_0^*, IV, II, good) over 0 as $k \equiv 1$ (resp. 2,3,4,5,0) modulo 6. One can show:

Proposition V.4.12: For these surfaces, we have the following:

k	over ∞	over k^{th} roots of 1	over 0	rank of Mordell-Weil group	Picard number
$6\ell+1$ $\ell \geq 0$	I_k	I_1	II^*	$r = \begin{cases} 4 & \text{if } \ell \equiv 4 \bmod 5 \\ 0 & \text{otherwise} \end{cases}$	$\rho = \begin{cases} k+1 \\ k+9 \end{cases}$
$6\ell+2$ $\ell \geq 0$	I_k	I_1	IV^*	$r = \begin{cases} 7 & \text{if } 3\ell+1 \equiv 0 \bmod 10 \\ 5 & \text{if } 3\ell+1 \equiv 5 \bmod 10 \\ 3 & \text{if } 3\ell+1 \equiv 2,4,6,8, \bmod 10 \\ 1 & \text{otherwise} \end{cases}$	$\rho = \begin{cases} k+14 \\ k+12 \\ k+10 \\ k+8 \end{cases}$
$6\ell+3$ $\ell \geq 0$	I_k	I_1	I_0^*	$r = \begin{cases} 6 & \text{if } \ell \equiv 2 \bmod 5 \\ 2 & \text{otherwise} \end{cases}$	$\rho = \begin{cases} k+11 \\ k+7 \end{cases}$
$6\ell+4$ $\ell \geq 0$	I_k	I_1	IV	$r = \begin{cases} 7 & \text{if } 3\ell+2 \equiv 0 \bmod 10 \\ 5 & \text{if } 3\ell+2 \equiv 5 \bmod 10 \\ 3 & \text{if } 3\ell+2 \equiv 2,4,6,8 \bmod 10 \\ 1 & \text{otherwise} \end{cases}$	$\rho = \begin{cases} k+10 \\ k+8 \\ k+6 \\ k+4 \end{cases}$
$6\ell+5$ $\ell \geq 0$	I_k	I_1	II	$r = \begin{cases} 4 & \text{if } \ell \equiv 0 \bmod 5 \\ 0 & \text{otherwise} \end{cases}$	$\rho = \begin{cases} k+5 \\ k+1 \end{cases}$
$6\ell+6$ $\ell \geq 0$	I_k	I_1	good, but cosingular	$r = \begin{cases} 9 & \text{if } \ell \equiv 9 \bmod 10 \\ 7 & \text{if } \ell \equiv 4 \bmod 10 \\ 5 & \text{if } \ell \equiv 1,3,5,7 \bmod 10 \\ 3 & \text{otherwise} \end{cases}$	$\rho = \begin{cases} k+10 \\ k+8 \\ k+6 \\ k+4 \end{cases}$

Note: $p_g = \ell$, $q = 0$ in all cases.

Proof: See Stiller [43]. □

APPENDIX I: THIRD ORDER DIFFERENTIAL EQUATIONS

In this appendix we wish to collect together a number of formulas and results on third order linear homogeneous differential equations with three regular singular points. Equations of this type arise in the examples in Part V. Some of the particular equations in those exmples were of the form

$$\left(\frac{d}{du} + R(u)\right)\left(\frac{d^2}{du^2} + \frac{1}{u}\frac{d}{du} + \frac{\frac{31}{144}u - \frac{1}{36}}{u^2(u-1)^2}\right)$$

on the u-sphere where the first order operator also has regular singular points at $u = 0, 1, \infty$ and no other singularities. It follows that

$$\frac{d}{du} + R(u) = \frac{d}{du} + \left(\frac{a}{u} + \frac{b}{u-1}\right) \qquad a, b \in \mathbb{C}$$

which has solutions of the form $cu^{-a}(u-1)^{-b}$, $c \in \mathbb{C}$, and that our third order equation is

$$\frac{d^3}{du^3} + \frac{(a+b+1)u - (a+1)}{u(u-1)}\frac{d^2}{du^2} + \frac{(a+b-1)u^2 + \left(\frac{319}{144} - 2a - b\right)u + \left(a - \frac{37}{36}\right)}{u^2(u-1)^2}\frac{d}{du}$$

$$+ \frac{\frac{31}{144}(a+b-3)u^2 + \left(\frac{47}{144} - \frac{35}{144}a - \frac{1}{36}b\right)u + \left(\frac{a}{36} - \frac{1}{18}\right)}{u^3(u-1)^3} \, .$$

For example, in one case appearing in Part V, we had $a = b = 3/2$ and the resulting equation was

$$\frac{d^3}{du^3} + \frac{4u - \frac{5}{2}}{u(u-1)}\frac{d^2}{du^2} + \frac{2u^2 - \frac{329}{144}u + \frac{17}{36}}{u^2(u-1)^2}\frac{d}{du} + \frac{-\frac{23}{288}u - \frac{1}{72}}{u^3(u-1)^3} \, .$$

At $u = 0$ the indicial equation is $v^3 - \frac{1}{2}v^2 - \frac{1}{36}v + \frac{1}{72}$ and the resulting exponents are $-1/6$, $1/6$, $1/2$. At $u = 1$ the indicial equation is $v^3 - \frac{3}{2} v^2 + \frac{11}{16} v - \frac{3}{32}$ and the resulting exponents are $1/4$, $1/2$, $3/4$. Finally at $u = \infty$ the indicial equation is $v^3 - v^2$ and the exponents are 0, 0, 1. Of course among the solutions of this third order equaiton are those of the second order equation

$$\frac{d^2}{du^2} + \frac{1}{u} \frac{d}{du} + \frac{\frac{31}{144} u - \frac{1}{36}}{u^2(u-1)^2}$$

which was discussed in Part V, §1. Note that this equation has exponent $\pm 1/6$ at $u = 0$, $1/4$, $3/4$ at $u = 1$, and $0,0$ at $u = \infty$.

Two important operations on our third order equations are changing parameter and "twisting". We record here the resulting formulas. If

1)
$$\frac{d^3}{du^3} + P \frac{d^2}{du^2} + Q \frac{d}{du} + R$$

is a third order equation and t is another parameter, then in terms of t the equation becomes

$$\frac{d^3}{dt^3} + P\left[\frac{du}{dt} + 3 \frac{\frac{d^2 t}{du^2}}{\left(\frac{dt}{du}\right)^2}\right]\frac{d^2}{dt^2} + \left[Q\left(\frac{du}{dt}\right)^2 + P \frac{\frac{d^2 t}{du^2}}{\left(\frac{dt}{du}\right)^3} + \frac{\frac{d^3 t}{du^3}}{\left(\frac{dt}{du}\right)^3}\right]\frac{d}{dt} + R\left(\frac{du}{dt}\right)^3 .$$

On the u-sphere, setting $t = 1/u$, this yields

$$\frac{d^3}{dt^3} + \left(P\left(-\frac{1}{t^2}\right) + \frac{6}{t}\right)\frac{d^2}{dt^2} + \left(Q\frac{1}{t^4} + P\left(-\frac{2}{t^3}\right) + \frac{6}{t^2}\right)\frac{d}{dt} + R\left(-\frac{1}{t^6}\right) .$$

Next we wish to find the equation whose solutions are g times those in 1) for some function g. This is the process we've called twisting. The result is

$$\frac{d^3}{du^3} + \left(P - 3\frac{g'}{g}\right)\frac{d^2}{du^2} + \left(Q - 2P\frac{g'}{g} - 3\frac{g''}{g} + 6\left(\frac{g'}{g}\right)^2\right)\frac{d}{du}$$

$$+ \left(R - Q\frac{g'}{g} - P\frac{g''}{g} - \frac{g'''}{g} + \frac{6g'g''}{g^2} + 2P\left(\frac{g'}{g}\right)^2 - 6\left(\frac{g'}{g}\right)^3\right).$$

For a general third order equation on the u-sphere to have regular singular points at $0, 1, \infty$ it must be of the form

$$\frac{d^3}{du^3} + \frac{\Lambda u + B}{u(u-1)}\frac{d^2}{du^2} + \frac{Cu^2 + Du + E}{u^2(u-1)^2}\frac{d}{du} + \frac{Fu^3 + Hu^2 + Iu + J}{u^3(u-1)^3}$$

which can also be put in the form

$$\frac{d^3}{du^3} + \left(\frac{a}{u} + \frac{b}{u-1}\right)\frac{d^2}{du^2} + \left(\frac{c}{u^2} + \frac{d}{u} + \frac{e}{(u-1)^2} + \frac{-d}{u-1}\right)\frac{d}{du}$$

$$+ \left(\frac{g}{u^3} + \frac{h}{u^2} + \frac{i}{u} + \frac{j}{(u-1)^3} + \frac{i-h}{(u-1)^2} + \frac{-i}{u-1}\right)$$

where

$a+b = A$	$a = -B$
$-a = B$	$b = A+B$
$c = E$	$c = E$
$-2c+d = D$	$d = D+2E$
$c-d+e = C$	$e = C+D+E$
$-g = J$	$g = -J$
$3g-h = I$	$h = -3J-I$
$-3g+3h-i = H$	$i = -6J-3I-H$
$g-2h+i+j = F$	$j = F+J+I+H$.

Note that such an equation depends on nine parameters and it has nine exponents determined locally. However, unlike the case of the classical hypergeometric function, we find that the local data does not determine the global properties. This can be seen by considering the indicial equations. The nine exponents completely determine a,b,c,d,e,g,j and $(i-2h)$, but cannot separate out i and h.

We have that at $u = 0$ the indicial equaiton is

$$v^3 + (a-3)v^2 + (c-a+2)v + g$$

at $u = 1$ it is

$$v^3 + (b-3)v^2 + (e-b+2)v + j$$

and at $u = \infty$ it is

$$v^3 + (3-a-b)v^2 + (2-a-b+c-d+e)v + (g+j+i-2h).$$

Finally, we consider the generalized hypergeometric function

$$_3F_2\binom{p_1,p_2,p_3;u}{q_1,q_2} =$$

$$1 + \left(\frac{p_1 \cdot p_2 \cdot p_3}{q_1 \cdot q_2 \cdot 1}\right) u + \ldots + \frac{p_1(p_1+1)\ldots(p_1+n)p_2\ldots(p_2+n)p_3\ldots(p_3+n)}{q_1(q_1+1)\ldots(q_1+n)q_2\ldots(q_2+n)(n+1)!} u^{n+1} + \ldots$$

which converges for $|u| < 1$ provided $q_1, q_2 \neq 0, -1, -2\ldots$. This function satisfies the third order equation

$$\frac{d^3}{du^3} + \frac{(p_1+p_2+p_3+3)u - (q_1+q_2+1)}{u(u-1)} \frac{d^2}{du^2} + \frac{(p_1p_2+p_2p_3+p_1p_3+p_1+p_2+p_3+1)u - q_1q_2}{u^2(u-1)}$$

$$+ \frac{p_1p_2p_3}{u^2(u-1)}.$$

At 0 the exponents are 0, $1-q_1$, $1-q_2$, at 1 the exonents are 0, 1, $q_1+q_2-p_1-p_2-p_3$, and at ∞ the exponents are p_1, p_2, p_3.

The function $_3G_2\left(\begin{matrix} p_1,p_2,p_3;u \\ q_1,q_2 \end{matrix}\right)$ is defined to be equal to $_3F_2\left(\begin{matrix} p_1,p_2,p_3 \\ q_1,q_2 \end{matrix}; 1-u\right)$. It also satisfies an easily computed third order equation.

BIBLIOGRAPHY

[1]   Bateman Manucript Project, Higher Transcendental Functions,
       McGraw-Hill Co., New York, 1953.

[2]   P. Bayer and J. Neukirch, "On Automorphic Forms and Hodge Theory,"
       Math. Ann. 257, 137-155 (1981).

[3]   D. Cox, "Mordell-Weil Groups of Elliptic Curves over $\mathbb{C}(t)$ with
       $p_g = 0$ or 1", Duke Math. J. 49 (1982), 677-689.

[4]   D. Cox and S. Zucker, "Intersection Numbers of Sections of Elliptic
       Surfaces," Inv. Math. 53, 1-44 (1979).

[5]   P. Deligne, "Equations Différentielles à Points Singuliers
       Réguliers," Lec. Notes in Math., 163, Springer-Verlag, Berlin-
       Heidelberg-New York (1970).

[6]   P. Deligne, "Travaux de Griffiths," Sem. Bourbaki 1969/70, no. 376,
       Lec. Notes in Math., 180, Springer-Verlag, Berlin-Heidelberg-New York
       (1971).

[7]   P. Deligne, "Théorie de Hodge I," Actes, Congres intern. math.,
       (1970).

[8]   P. Deligne, "Théorie de Hodge," Publ. Math. I.H.E.S., 40 (1971).

[9]   M. Eichler, "Eine Verallgemeinerung der Abelschen Integrale," Math.
       Z., 67 (1957), 267-298.

[10]   P. Griffiths, "Differential Equations on Algebraic Varieties,"
       Princeton lectures, unpublished.

[11]   R. Hartshorne, "On the De Rham cohomology of Algebraic Varieties,"
       Publ. Math. I.H.E.S., 45 (1975).

[12]   W. Hoyt, "On Elliptic Surfaces and Automorphic Forms," preprint.

[13]   W. Hoyt, "An inhomogeneous Riemann-Roch theorem for generalized
       automorphic forms," preprint.

[14]   E. Ince, "Ordinary Differential Equations," Dover Publications,
       New York (1956).

[15]   N. Katz and T. Oda, "On the differentiation of the De Rham cohomology
       classes with respect to a parameter," J. Math. Kyoto Univ., 8-2
       (1968), 199-213.

[16]   K. Kodaira, "On compact analytic surfaces II," Ann. of Math., 77
       (1963), 563-626.

[17]   K. Kodaira, "On compact analytic surfaces III," Ann. of Math., 78
       (1963), 1-40.

[18] S. Lang, "Introduction to Modular Forms," Springer-Verlag, Berlin-Heidelberg-New York (1976).

[19] S. Lang, "Elliptic Functions," Addison-Wesley, Reading (1974).

[20] J. Manin, "Algebraic curves over fields with differentiation," AMS Trans., 50 (1966).

[21] J. Manin, "Rational points of algebraic curves over function fields," AMS Trans. (2), 37, pp. 59-78.

[22] E. Picard, "Traité D'Analyse," Tome III, Gauthier-Villars, Paris (1908).

[23] H. Poincaré, "Sur les Groupes des Équations Linéaires," Oeuvres, Tome II, Gauthier-Villars, Paris (1916), 300-401.

[24] T. Sasai, "Monodromy representations of homology of certain elliptic surfaces," J. Math. Soc. Japan, Vol. 26, No. 2 (1974), 296-305.

[25] G. Shimura, "Introduction to the Arithmetic Theory of Automorphic Forms," Princeton University Press, Princeton (1971).

[26] G. Shimura, "Sur les intégrales attachées aux formes automorphes," J. Math. Soc. Japan, Vol. 11, No. 4 (1959).

[27] T. Shioda, "On elliptic modular surfaces," J. Math. Soc. Japan, Vol. 24, No. 1 (1972), 20-59.

[28] P. Stiller, "Differential equations associated with elliptic surfaces," J. Math. Soc. Japan, Vol. 32, No. 2 (1981).

[29] P. Stiller, "Elliptic curves over function fields and the Picard number," Amer. Jour. of Math., Vol. 102, No. 4 (1980).

[30] P. Stiller, "Monodromy and invariants of elliptic surfaces," Pac. Jour. of Math., Vol. 92, No. 2 (1981).

[31] P. Stiller, "A note on automorphic forms of weight one and weight three," preprint.

[32] P. Stiller, "A generalized Shimura isomorphism," preprint.

[33] A. Weil, "Généralization des fonctions abéliennes," J. Math. Pures Appl., 17 (1938), 47-87.

[34] U. Schmickler-Hirzebruch, "Elliptische Flächen Über $\mathbb{P}^1_{\mathbb{C}}$ mit Drei Ausnahmefasern und die Hypergeometrische Differentialgleichung" Diplomarbeit, Universität Bonn, 1978.

[35] N. Ja. Vilenkin, "Special Functions and the Theory of Group
      Representations," translated by V. N. Singh, AMS, Providence (1968).

[36] P. Stiller, "Special Values of Dirichlet Series, Mnodromy, and the
      Periods of Automorphic Forms," I.H.E.S. preprint.

[37] B. Gross, "On the periods of Abelian Integrals and a Formula of Chowla
      and Selberg", Inv. Math. 45 (1978).

[38] S. Zucker, "Generalized Intermediate Jacobians and the theorem on
      normal functions", Inv. Math. 33 (1976).

[39] I. S. Gradshteyn and I. M. Ryzhik, "Table of Integrals, Series, and
      Products", Academic Press, New York (1965).

[40] H. Exton, "Handbook of Hypergeometric Integrals", Ellis Horwood
      Limited, Chichester (1978).

[41] M. Atiyah and W.V.D. Hodge, "Integrals of the second kind on an
      algebraic variety", Annals of Math., Vol. 62 (1955).

[42] P. Deligne, "Valeurs de fonctions L et périodes d'intégrales",
      313-346 in "Automorphic Forms, Representations, and L-Functons
      (Part 2)", Proc. Sym. Pure. Math. 33, AMS. Providence (1979).

[43] P. Stiller, "The Picard numbers of some elliptic surfaces", preprint.

[44] S. Chowla and B. Gross, "An integral formula", unpublished.

[45] J. Manin, "Periods of parabolic forms and p-adic Hecke series,"
      Math. USSR Sbornik, Vol. 21 (1973), no. 3.

[46] M. Razar, "Values of Dirichlet series at integers in the critical
      strip," in Modular Functions of One Variable VI, Springer Lecture
      Notes 627, Springer-Verlag (1977).